Captain Kevin 'Sully' Sullivan has made flying his passion and his life for the past 40 years. He graduated in 1977 from the University of Colorado with a Bachelor of Science in Aerospace Engineering and earned his FAA Private Pilot Licence there before pursuing a career in the United States Navy. Designated a Naval Aviator in 1978, he was transferred to Naval Air Station Miramar (Fightertown) to fly the F-14 Tomcat in 1980. He was deployed to the Indian Ocean onboard USS *America* and USS *Enterprise* while assigned to Fighter Squadron 114 (VF-114 Fighting Aardvarks), and was chosen to attend the Navy Fighter Weapons School (TOP GUN).

In 1983 he was selected as the first US Navy Exchange Pilot to the Royal Australian Air Force, in the role of a Fighter Combat Instructor flying the Mirage 3. He joined Qantas Airways in 1986 and flew the Boeing 747 and 767 before transitioning to the Airbus A330 in 2004.

As Captain of Qantas Flight 72 (QF72) between Singapore and Perth, Western Australia, on 7 October 2008, he narrowly averted a horrific air disaster when a fault in the plane's automation caused the plane to suddenly pitch down twice. He was medically retired in 2016.

# NO MAN'S LAND

Kevin Sullivan

ABC BOOKS

First published in Australia in 2019
by HarperCollins*Publishers* Australia Pty Limited
ABN 36 009 913 517
harpercollins.com.au



**HarperCollins*Publishers***
Level 13, 201 Elizabeth Street, Sydney, NSW 2000, Australia
Unit D1, 63 Apollo Drive, Rosedale, Auckland 0632, New Zealand
A 53, Sector 57, Noida, UP, India
1 London Bridge Street, London, SE1 9GF, United Kingdom
Bay Adelaide Centre, East Tower, 22 Adelaide Street West, 41st Floor, Toronto, Ontario, M5H 4E3, Canada
195 Broadway, New York, NY 10007, USA

A catalogue record for this book is available
from the National Library of Australia

ISBN: 978 0 7333 3974 5 (paperback)
ISBN: 978 1 4607 1090 6 (ebook)

Cover design by HarperCollins Design Studio
Cover image by Louie Douvis/Fairfax Syndication
Typeset in Adobe Caslon Pro by Kirby Jones
Printed and bound in Australia by McPherson's Printing Group
The papers used by HarperCollins in the manufacture of this book are a natural, recyclable product made from wood grown in sustainable plantation forests. The fibre source and manufacturing processes meet recognised international environmental standards, and carry certification.

'One day you will ask me which is more important?
My life or yours? I will say mine and you will walk away not
knowing that you are my life.' **Kahlil Gibran**

*For Lauren*

# CONTENTS

# PROLOGUE

'Sully … Carrot. One versus one, neutral set-up. Angels fifteen. Four hundred knots. Call ready.'

Abeam my Tomcat F-14 is a blue and grey camouflaged Skyhawk A-4. The pilot's callsign is 'Carrot' because of his full head of orange hair. He's an extremely lethal adversary pilot who flew fighters in the US Marine Corps before his personal invitation to become a Top Gun instructor. His exemplary performance during his course earmarked him for personal selection to this prestigious position.

My callsign is 'Sully', coincidentally matching the nickname of pilot Chesley Sullenberger, who in 2008 famously performed an emergency landing of US Airways Flight 1549.)

It's 1982 and the height of the Cold War. I'm a fighter pilot in the US Navy and have been selected to attend Top Gun, five weeks of 'postgraduate'-level training in air-to-air combat and tactics, formally known as the US Navy Fighter Weapons School. Few fighter pilots have the opportunity to attend this school – in fact, a squadron may only be allowed to send one two-man crew per year.

The ground instruction phase of the Top Gun training

provides us with current intelligence on potential threat aircraft and weapons systems. Because we can sail into harm's way at the drop of a hat, the flight phase has to prepare the crew to operate in a one-versus-unknown environment. If we ever fly in combat over Europe, we'll be fighting against an enemy with inferior aircraft but superior numbers. The sobering message from the course is that we have to learn to survive and fight in an arena of one-versus-four at a minimum.

On completion of the course, we will be awarded the coveted Top Gun patch, which adorns a special position on a pilot's flight suit, just below the left shoulder. It's the PhD of fighter qualifications.

The Top Gun course takes a building-block approach, starting with the one-versus-one scenario: two aerial gladiators facing off to the death. The opponent, nicknamed the 'bogey', is a Top Gun instructor hand-picked for their skill and lethality. Every engagement involves brutally employing the Top Gun motto: 'No Points for Second Place'.

On this day, flying above the Pacific Ocean far from our base in San Diego, I'm progressing through the one-versus-one phase. I'm flying my F-14 Tomcat versus a dissimilar aircraft that mimics a Cold War adversary, in this case, an A-4 Skyhawk (sometimes it's an F-5 Tiger 2 fighter). The instructors flying these jets simulate known Soviet tactics. For the students, it's hard work just to survive; the bogey pilots are the best in the navy.

The A-4F Skyhawk is a single-seat jet that the school has highly modified to have the same, if not superior, thrust-to-weight performance as the F-14, but with an inferior weapons system and missile capability. We've been instructed on the ways to fight and defeat this threat. Basically, the Tomcat's pilot has to use its superior thrust and turning capability to force the fight into the vertical. The pilot needs to maintain his energy, in the form of speed and use of g-force, to succeed.

But aerial combat is much more than just pulling maximum g-force and shooting down your opponent. It's a three-dimensional chess match of energy management, weapons-system employment, communication, teamwork and endurance of the physical exertion of flying to the aircraft's maximum g capabilities. Plan A is rarely executed as desired – the high-speed, high-stakes arena of air-to-air combat requires continual innovation, creativity and discipline to survive and to win. I feel sure I am prepared. I have demonstrated my skill and aggression flying with my squadron. They know I am ready to advance my lethal skills to the next level.

Carrot has set the starting parameters for our first visual engagement. Our altitude is 15,000 feet, separated by a distance of 5000 feet laterally, and we will start the fight at 400 nautical miles per hour of airspeed (our speed relative to the air). Now it's time to fight and to perform.

'Sully's ready,' I say as I select full afterburner on my engines. The fuel flow through the engines is a staggering 48,000 pounds per hour.

'Fight's on!'

'Fight's on,' I repeat as I turn hard into the Skyhawk A-4, hearing the wing roar as the big Tomcat F-14 wing slices through the air. The g-force escalates to six and a half times more than I normally weigh, and I have to twist and contort my body to maintain sight of the A-4. I'm exerting myself as much if I were wrestling an elephant. We are turning at the maximum rate, and my objective is to make Carrot trade his airspeed for turning advantage. He'll lose energy if he does this, and I'll maintain mine, then I'll have the energy advantage to vertically separate from him and shoot my 'missiles'.

This is the plan. I have to use my airspeed, altitude and g in order to maintain my energy state, and I have to compare that with the aircraft I'm fighting against. If I observe that he's turning tighter, I know he's trading energy for positional advantage. I must be patient while I turn hard to neutralise his threat. I'm seeing exactly what I want to see, and watch the A-4 shoot past my tail. We started in a neutral position and now he's almost behind me; therefore he's traded all his energy for this advantage of position … right? Wrong. As Carrot slips past my tail, I roll my wings level and pull straight up. Surely there's no way he could match my climb – but he does.

I got it all wrong. Carrot was climbing better than me. He maintained his energy and tricked me into thinking he was out of puff. Now I'm paying the price as he latches onto my tail.

'Guns, guns, guns,' he calls, simulating an attack on my highly energised but disadvantaged Tomcat.

Breathing hard from the high g-forces, I shake my head in disgust. 'Knock it off,' I yell. It's our agreed safety phrase to stop fighting, to maintain sight of each other and set up for the next exercise.

'Let's not do that again, Sully,' says Breeze, my radar intercept officer. Breeze is from Virginia, but he's adapted quickly to the 'breezy' lifestyle of southern California. When he isn't in my back seat fighting a Top Gun instructor, he likes to surf, sit in the sun and play racquetball. He's known as a bit of a wild child and we will later share a stateroom on the *USS Enterprise* when the squadron deploys for cruise. He knows his stuff and is up to the challenges of aerial combat.

I only have a minute or two to figure out how to not 'do that again' before we are ready for the next fight. I just got my ass kicked. On my kneepad I quickly scribble some shorthand notes depicting what each aircraft did in sequence. We also carry small tape-recorders that we use to help reconstruct the fights later.

My energy management is good, but Carrot's is better. I have to do something different to gain an advantage or I'll be eating Carrot's machine-gun shells again. Oh yeah, at this level, it is all very real and very serious.

The A-4 cockpit isn't as big or open as the bubble canopy of the F-14. I can easily pivot my body to look back between the tails, but the A-4 pilot can't. It's the same visual restriction that a MiG pilot would have: the bad-guy aircraft Carrot is simulating. As long as he has me in front, he has the advantage. I need to try something different. I aim to exploit the reduced rearward visibility of the A-4, keeping in mind that the F-14 is one big jet. Making the bogey lose sight of me might give me the separation I need to jab the bogey from a distance with my superior air-to-air missiles. But this plan will test the limits of the coordination between Breeze and me; two sets of eyes are better than one.

I brief Breeze on my plan so he's prepared for my abrupt handling of our jet.

'Okay, just don't get us shot this time!' he barks.

We're almost back at altitude to start the next engagement. This time, Carrot has positioned his A-4 closer to me. Our starting speed is slower – 300 knots – a disadvantage to me and an advantage to the A-4. We're ready.

'Fight's on!'

Even though this is training and I see an A-4, I imagine a potential future reality: an enemy aircraft about to attack from close range, and I have to neutralise his advantage and go on the offensive.

In full afterburner, I turn at the maximum rate towards the A-4. Carrot is doing the same. As we pass canopy-to-canopy, I reverse my turn to fake a change of direction. I can almost see Carrot grinning at this apparently fatal move, but when I pass behind his aircraft I reverse my turn again and dive.

My plan works.

'No joy,' Carrot transmits to indicate he's lost sight of me. Unfortunately, I've lost sight of him, too.

But Breeze has not. He confirms that he can still see Carrot, so we can continue the fight safely. I've banked my F-14 on one wing and am pushing the nose down slightly to accelerate without any g-force on the wing. We call this 'unloading' and we are lightly elevated out of our ejection seats. The afterburners are quickly adding energy to my unloaded Tomcat.

Breeze is pretzelled around, pushing against a small handrail on his instrument panel while trying to talk my eyes onto the A-4. I glanced at my airspeed gauge: 350 to 400 then 450 knots. As we build speed, we're building the separation I need to attack rapidly – if only I could see my target which is rapidly disappearing behind our wing and tail.

Breeze is getting frustrated. Using a clock-code relative to our aircraft – twelve o'clock is straight ahead, six o'clock is directly behind etcetera – he's trying to help me regain sight of Carrot, but I still can't see him because we've opened up some distance with my acceleration. Luckily, Carrot still can't see me either.

'No joy,' I reply when Breeze makes a clock-code sighting.

This happens a few more times in rapid succession with the same result. In exasperation, Breeze jabs his finger on the canopy. 'He's right there!'

Crazily enough, that works. I confirm my sighting with Breeze. Time to attack.

I reduce my power slightly as I turn at maximum g to point at Carrot. I instruct Breeze to activate one of the short-range radar modes, that he controls in his cockpit, to provide a radar lock on the A-4.

Now I can start shooting my missiles.

'Fox 1, A-4, left-hand turn, angels 15.' This indicates I've launched a radar-guided Sparrow missile at the A-4, which is turning left at 15,000 feet.

'Continue,' Carrot grunt, as he racks his jet into a maximum g defensive turn. He still can't see me or my missile.

I launch another Sparrow missile his way and make the notional call.

We're closing rapidly on Carrot. I move my thumb down one notch on my control stick and select the shorter-range Sidewinder missile, its guidance based on heat emanating from the target. Through my headphones, I hear the seeker head on my Sidewinder start barking its signal; it has locked on to the heat plume coming from the A-4's engine.

I pull the control stick trigger and send a Sidewinder towards Carrot's wildly manoeuvring jet. 'Fox 2, A-4, right-hand turn, angels 13.'

Again I hear Carrot's grunting reply: 'Continue.'

Okay. I launch another Sidewinder and make the call for my shot. If this was a real fight with real missiles, the enemy would have been in flames after the Sparrow missile I'd shot seconds before. He wouldn't have seen it coming – we don't give the enemy a heads-up that death is streaking his way.

We're getting closer. My thumb makes another selection to arm my gun: a 20-millimetre Vulcan Gatling gun, with six rotating barrels capable of firing 6000 rounds per minute. I decide to use it now to finish this fight. The shape of the A-4 is filling my forward windscreen as I manoeuvre to put the gunsight on my target.

Just before I pull the trigger, Carrot finally sees me and turns aggressively to defeat my gunshot. I pick a point in front of his aircraft and pull the trigger, sending a stream of projectiles into his path. He flies straight through them.

'Snapshot!' I say to indicate a high-angle gun attack.

I keep pointing at his aircraft and reselected afterburner. Then I escape at supersonic speed.

'Knock it off,' Carrot transmits. The fight is over. He couldn't catch my accelerating F-14.

I confirm the end of the fight and add another call.

'Winchester' means I've fired all my imaginary missiles and am out of ammunition.

I train like I fight, and in this fight I used everything I had to survive and win.

*

That day was a turning point in my Top Gun training and my military aviation career. I improvised and executed a complex attack against a worthy adversary, fired all my missiles, and effectively communicated and coordinated with my radar intercept officer.

Speed is life. If you have it then you can manoeuvre, and if you can manoeuvre then you can survive. In our world of one-versus-many, you can't afford to slow down.

We graduated from Top Gun four weeks later, and Breeze and I were awarded the coveted patch to sew onto our flight suits. I was full of confidence, ready to handle any combat scenario that was thrown at me. It didn't matter if there was only one of me and many of them; I'd win because, in war, there are no points for second place.

Back then, my country and a large portion of the Western world were fighting a Cold War. Although no shots were fired and no battles fought, we were on the verge of conflict at the push of a button.

As it turned out, I was lucky, and I survived those years of conflict without having to employ my deadly skills as a fighter pilot. I didn't take a life and I didn't fire a shot, but I did lose many good friends.

Eventually, I was invited to share my lethal experience with the Royal Australian Air Force as an exchange officer. My military career spanned nine years. I learned to fight and survive in the most extreme arenas of military aviation. Then I traded my navy dress blue uniform for that of a commercial pilot. Years later, on 7 October 2008, I went into battle in this uniform, using all my hard-won skills to survive a conflict that I was forced to fight by circumstance. I wasn't fighting for my country or a way of life, but for my own life and the lives of those entrusted to me on Qantas Flight 72 from Singapore to Perth. We were in an Airbus A330, equipped with the latest in automation technology. Unfortunately, that technology, which was designed to keep us safe, dramatically malfunctioned and harmed us.

In aviation, seconds count. Decisions made in extremis, unforeseen circumstances or the hand of Fate can alter one's present and future existence drastically and sometimes fatally. And in just two seconds, my life and the lives of the other crew and 303 passengers aboard QF72 changed forever.

To land that plane, I had to fall back on my fighter-pilot experience and the skills I'd been taught by my instructors and honed during my five weeks at Top Gun. I'm here to tell you that I won the fight.

This is a good-news story. It's a success story. We all survived and so did the plane. But in those two seconds, when the plane's automation malfunctioned, we were brutally thrust into the unchartered territory of No Man's Land. Our fight continues.

# I

# KNIFE FIGHT

# 1.

'Geez, Kev,' says Pete. 'I thought about going sick on this trip.' He points at the amber master caution light on the console. The plane's computers have detected a fault in the air system connected to one of our engines.

Behind us, Ross chuckles. My reputation as a 'shit magnet' is widespread throughout the airline. This is aviation slang for a pilot who has attracted more than their fair share of malfunctions and emergency situations. It is true, though. As a Qantas pilot, I've had to deal with more shit than most. But they also know my reputation for being cooler than a cucumber smoothie when Fate tests me.

'Yeah, well, that better be the worst thing that happens today.' I stare them down with one eyebrow raised.

Good thing these guys are two of the best. First Officer Pete Lipsett has had an extensive career flying helicopters in the Royal Australian Navy. He's a few years younger than me, and I'm happy to have his experience next to me on the flight deck. Second Officer Ross Hales has been at Qantas for less than twelve months and on the Airbus A330 fleet for six months.

His flying experience was built in general aviation, flying throughout Western Australia and the Northern Territory, and he's a mature and calm operator. I would hate to face off against him on the footy field. He's a strapping young lad.

Like me, these pilots are in the career of their dreams and enjoying their passion for aviation. They also know I calmly handle everything that's thrown my way, but they can't suppress their Aussie sarcasm.

We all laugh, then get back to business.

It's 7 October 2008, and for me, the captain of Qantas Flight 72, life is at its peak. I'm fifty-three years old, a captain in a profitable international airline with a good salary, and continuing my passion for flying as a commercial pilot.

*

The day before, I flew with Pete and Ross from Brisbane to Singapore. We'd never flown together before this trip; such is the nature of commercial flying. However, our company's standardised approach to procedure ensures we work efficiently together regardless and we immediately got along. We shared some satays and a beer or two the night before our flight, and this helped to solidify our team. Is there anything beer *can't* do?

We were up at eight and met in the hotel lobby, relaxed and well rested. I nominated myself to fly the first sector down to Perth. It was going to be a long-duty day as we were scheduled to return to Singapore that evening. Ross followed my lead and we left our luggage at the hotel.

In our staff minibus on the way to the airport, I walked down the aisle introducing myself as the captain and shaking hands with my cabin crew. There was a familiar face or two, but for the most part I hadn't flown with them before.

I do this before every flight: make contact with the crew so they can see who I am. I'm in a leadership position and I'm forming my team – after all, if the shit hits the fan, I want to see who's going to open the doors in an evacuation. I also instruct them as a group that 'no one gets hurt today', as I brief them on any expected turbulence encounters. I'm responsible for their safety, and I reinforce that responsibility by stating I'll do my part to ensure we all walk off at the end of the flight intact.

On the minibus I introduced myself to my Sydney-based flight attendants Kimberly, Rory, Alia and Jill, and to Lisa, the customer service manager.

I asked Rory who he'd bribed to be rostered with four lovely ladies. In his South Australian twang he said they were lucky to be flying with a good bloke like him. The ladies unleashed some playful abuse in response causing Rory to slide down in his seat with a smirk.

Lisa and I had flown together a year or two before. Recently, she'd been promoted to her current supervisory position. She was very professional during our briefing and we refrained from slipping into any banter. We talked about the particulars of our sector to Perth, including the flight time and any special handling requirements she might have. She advised me that the passenger load was full – in fact, it was oversold in economy class – but there would probably be some Qantas staff waiting at the airport check-in area, trying to get home.

'I've got a note from the airport staff. Lots of staff travellers trying to get back to Oz. Some have been here all week trying to get out,' Lisa said.

'Ah, the joys of staff travel. Can we help them?' I asked.

'Of course. Our crew rest seats are available and I will manage it once we get airborne.'

We'd do our best to get a few of them on our flight, and the offer of the crew's rest seats would allow us to do just that.

Sitting in the rear of the bus were four crew members from the Auckland base. Although they wore the Qantas uniform, they were employed by Jetconnect, a wholly owned Qantas subsidiary company based in New Zealand.

I met Fuzzy Maiava, a senior flight attendant from Auckland who was built like a rugby forward. He had an infectious smile and warmly greeted me with a crushing handshake. Softly spoken, he's the type of person you like immediately. With him were Samantha, Tasha and Jen, three young Auckland-based flight attendants who meekly shook my offered hand. They were in their early twenties and looked so young to me.

I finished my briefing and returned to my seat. My team was formed; nine cabin crew and three pilots, relaxed and ready to work.

As expected, the airport check-in area was buzzing with passengers, and a large group of airline staff members had gathered near the standby desk, trying to get home. As we entered, they gave us anxious looks. Some had been trying to leave Singapore for several days, but the flights had been full without much room for them travelling on their low-priority standby tickets. These travel experiences took the glitter off the perceived benefits of cheap travel for airline employees; these people were all desperate to get home to Australia. As planned, I asked Lisa to coordinate with the airport staff to take some of them with us.

Ross, Pete and I made our way to the briefing office to prepare for the flight. It would take four hours and forty-five minutes, the weather was forecast to be fine for our arrival, and I noted the aircraft's rego letters: 'VH-QPA'.

The three of us decided, on my recommendation, to order some extra fuel to cover any unforeseen events once we reached Perth. Aviation is all about managing risk, and I always treat this airport as isolated because there are no others in close vicinity.

All of our flights are required to be planned with suitable contingency airports. Our flight plan used Learmonth, a shared military and general aviation airfield, for mandatory point-of-no-return planning. It provides a point between the two airports to permit a divert decision if Perth weather or airport operational issues prevent us from proceeding. The extra fuel provides extra time to deal with anything out of the ordinary.

With the flight plan safely stowed in Ross's bag, the three of us headed to our aircraft. The A330 was buzzing with activity as we entered the cabin. The crew were inspecting their safety equipment and stowing the heavy metal food carts in the galleys. Before heading to the cockpit, I got an update from Lisa about the extra staff passengers and gave her clearance to start boarding the aircraft once they were ready.

In the cockpit, I took my place in the left seat. Pete, Ross and I performed our usual procedures to prepare the aircraft for flight.

Our workplace is unique; it's highly automated and deceptively simple in its design. The cockpit instrument and side panels are coloured a light blue. Directly in front of me are the two main electronic flight displays, which are repeated on the right for the first officer, and in the centre are two screens to display the engine instruments (upper) and the various aircraft systems (lower).

The left display, known as the primary flight display, projects the basic performance instrumentation a pilot needs to fly and control the aircraft. In the centre is the artificial horizon that bisects the display. Superimposed are the horizontal hash marks depicting nose attitude above and below the horizon in five degree increments. The sky is coloured blue and brown represents the ground. As the aircraft climbs and descends, the amount of blue and brown changes accordingly, and the pilot can reference the angle of the nose in relation to the horizon using the attitude markings. Superimposed across the centre of the display are the artificial yellow wings of the aircraft. The vertical tapes on either side are the speed (left) and altitude (right).

Inboard from the primary display is the navigation display. It's a little like your car's map and GPS system, and it projects a god's-eye view of the aircraft's path along the flight plan from departure point to destination. It also displays weather radar information and ground-based navigation aids. (These integrated displays have replaced the round instrument dials used in earlier-generation aircraft.)

Across the top of this array is a slim instrument panel housing a set of control buttons and selector knobs. These are used by the pilot to set speed, heading and altitude targets on the primary display when flying manually, and also to engage either of the aircraft's two autopilots. With an autopilot engaged, these selector knobs allow the pilot to control the aircraft's performance in speed, heading and altitude.

This panel also contains the master caution and master warning lights. These may look like small unassuming recessed pushbuttons, but they are critically important. The master caution light and switch will glow amber to alert us to a system fault. The master warning will illuminate red for more significant failures of equipment and critical systems. We hope we don't see their coloured lights at any time today.

The centre instrument panels contain the engine thrust (or power) levers, the engine fuel controls, the flaps and the speed-brake handle. On either side, close to the pilot's inboard knee, are the computer keyboards used to key in navigation and performance figures, as well as independent radio control panels.

Between the thrust levers and the keyboards are the two large trim wheels. Conventionally, as an aircraft increases speed, the pitch (or nose) of the plane wants to move up. The pilot uses a small switch on the control yoke to trim the tail down to counteract this force. The opposite is true for a speed reduction. It's like sailors on a boat, trimming the sails to relieve rudder pressure for the skipper.

Airbus has automated this function and the tail is automatically trimmed as the aircraft changes speed. The trim wheels are silently working to keep the aircraft in trim. These wheels can also be moved manually if the auto function is not available.

Above the pilot's head are the aircraft's systems control panels. All the support systems live here, such as fuel, air-conditioning, pressurisation and hydraulics. Their control switches are very simple – on, off or auto – and these push-button switches disguise the high level of automation and computer control that governs their operation.

On either side of the pilot's seat is a control sidestick. This has replaced the traditional control 'yoke', and is used by the pilot for manual control of the aircraft through the aircraft's fly-by-wire control system. This is the next generation of flight control. The pilot's control stick or column is no longer mechanically connected to the control surfaces on the wings or tail; rather, it's electrically connected to dedicated flight-control computers that move the controls in response to the pilot's requested input. The pilot, when flying the aircraft with or without the autopilot engaged, is interacting directly with a computer programmed to keep the plane operating safely.

There are three primary flight control computers (PRIMs) and two secondary computers (SECs) that make up the electronic flight control system. The overhead panel houses the control switches for the PRIMs and the SECs, while the actual computers are stored under the floor of the cockpit.

A significant safety enhancement to the Airbus is the use of flight envelope protection. In simple terms, the flight control computers are programmed to prevent the pilot from flying too fast or too slow, with too much angle of attack or with too much angle of bank when turning. With all computers operational, the aircraft is fully protected and this state is termed Normal Law. No matter how hard the pilot pushes or pulls, Normal Law will keep the aircraft safe. Also, pre-programmed manoeuvres will activate autonomously for excessive speed or angle of attack. These are the flight envelope protection modes.

If there is degradation to the PRIM's operation, the flight control system reverts to an Alternate Law of operation and some or all of the protections are lost. Alternate Law is not an emergency condition, but the pilot is alerted to this state change through the aircraft's warning system.

The aircraft's computer systems need to know where the aircraft is in space: its altitude and speed, the temperature of the air mass it is operating through and its position relative to the ground below and the aircraft's body angle relative to the air it is moving through (angle of attack). This spatial position is provided through the three Air Data and Inertial Reference Units; the ADIRUs.

Raw data sent into the ADIRU comes from the external sensors and probes protruding from the forward area of the aircraft's nose, and from GPS- and gyro-supplied latitude and longitude. What comes out are digitised values of these parameters that the PRIMs and SECs use to maintain the aircraft in safe flight. The computer-controlled support systems such as pressurisation, air conditioning, engine thrust, also rely on these digital ADIRU values for their normal operation.

Any corruption to this data, usually through an internal blockage or damage to the probes, can force the flight control system to revert from Normal Law to Alternate Law. The logic is that the computers cannot protect the aircraft if they don't know where it sits in the air mass, so the protections are removed. The plane still flies normally and there is always one flight control computer available to service the pilot's commands.

All of these computers are housed underneath the cockpit in an equipment area, and their operation is transparent to the pilots. We don't normally access this area while the aircraft is in the air but our ground engineers often do between flights.

This all sounds a lot like science fiction. It might be hard to get your head around, but these days, aeroplanes, are controlled by a version of 'HAL', the sentient computer in *2001: A Space Odyssey*. This automation is advertised by Airbus as an enhancement to safety, and they assure pilots and the public that there are enough back-up systems in place to cater for a wide range of failure situations. There is never any suggestion that the automation can fail and pilots do not train to address this abnormality.

On QF72, our preflight preparation ensured that all the automated systems were activated and operational, and we interacted with the computer keyboards to insert the navigation and performance figures from our flight plan. How easy could it be?

# 2.

The time is 0925. We're a bit late after taking time to seat our full load of passengers, but we'll easily manage that. There's no spare seats on our A330.

'Love your work, Kev,' Pete quips, and I call for the preflight checklist.

We start the engines as we begin our push back from the gate. All is going smoothly, and we're relaxed as we receive our taxi clearance from Singapore Ground Control and then our take-off clearance from the air traffic control tower. This has to be my fastest start-to-take-off sequence in history. The airport seems deserted and, after eight minutes we're on the roll for take-off.

Our climb out of Singapore and over the Straits of Malacca is serene. I have flown this route many times in the past. It is a rare treat to fly it in daylight and the topography of the Indonesian archipelago and its numerous islands are on display. The waters of the Malacca Straight are churned up by the propellers of every variety of ship imaginable, from freighters and cruise ships to military vessels and leisure craft. It is the busiest waterway on the planet and I must stop myself from looking down for too long as we set our course south.

I'm still manually flying our preloaded departure, using my instrumentation on my main display to follow steering commands. It looks like a yellow cross superimposed on the blue and brown background, which depict the sky and the ground with an artificial horizon in between. My adjacent navigation display provides a magenta-coloured line as another form of guidance as we fly the departure procedure loaded into our navigation computer.

Minutes after our take-off, the serenity of our departure is suddenly broken by a loud *ding* from our flight-deck speakers. Our eyes fix on the amber-coloured master caution light illuminated on the instrument panel above our flight displays. I reach up and push a button as I say to Ross and Pete, 'Engaging autopilot 1.'

Pete has already cancelled the master caution alert by pushing on the square button connected to this indicator light. He confirms my autopilot connection and begins the procedural reconfiguration sequence that was drilled into us during our Airbus training course.

The A330 entered service in 1994 and is the twin-engined version of the four-engined A340 that first flew in 1991. They share systems commonality with only minor differences in their operation.

I started my Airbus A330 conversion course in December 2004. The course took about four months to complete. It is written and structured by Airbus and supplied to airlines as part of their purchase package. The flight simulator, essentially a box mounted on hydraulic jacks, is used to familiarise the pilots with the cockpit and aircraft systems. It contains a full flight deck with interactive functionality so we can explore the flying characteristics and systems operation without leaving the ground.

A new method of reconfiguring systems faults and failures was introduced and practised extensively. The automated systems identify a fault through the master caution or warning lights and associated aural chimes. A message is displayed on the top screen of the centre instrument panel and the reconfiguration procedure is also displayed for the pilots to follow. This replaces the paper checklist procedure on previous generation aircraft and helps to reduce pilot workload. As designated switches are cycled and reset, the fault sequence is updated until the procedure is completed.

A fault message must 'latch' or remain displayed for this procedure to be initiated and completed. It is unusual for a fault to not latch but flash momentarily, indicating a transient condition that clears itself.

Each support system on the Airbus is controlled by a computer that monitors the routine operation of that system and alerts the pilots to any errors or faults. It's then up to us to rectify the fault by following the sequence of actions displayed on the two screens in the centre of our instrument panel. This sequence can generate a breakdown of flight-path monitoring, and it's up to the pilots to exercise a special discipline to maintain a safe operation.

This is the discipline of 'Aviate, Navigate, Communicate' taught to every fledging aviator, and I learned my trade in the US Navy.

Within weeks of graduating from university, I received my orders to commence flight training in Pensacola, Florida. The first phase involved classroom academics, survival and physical training. At the end of the two-month ground syllabus, we received orders to report to a basic training squadron located nearby. I was to fly the T-28 Trojan: a fully aerobatic carrier-capable trainer with a single radial-engine that produced 1000 horsepower of thrust. This type of performance could easily match most early World War II fighter aircraft – quite a handful for a neophyte aviator.

I was ordered to Naval Air Station Whiting Field, just north of Pensacola. This was the home of VT-4, Training Squadron 4, and I was assigned a primary flight instructor, Lieutenant Junior Grade Martin 'Ace' Powers.

I met Ace in August 1977, and his mission was to prepare me for my future in naval aviation. Ace, whom I addressed as 'sir', was a few years older than me but had been flying out in the fleet for the past three years. His brown leather flight jacket was crinkled and worn, unlike my new, shiny one that was hard for me to remove because I felt so good wearing it. He had flown helicopters from the decks of small destroyers, and he'd seen an abundance of black nights, foul weather and pitching seas while flying in the North Atlantic. He was a fleet veteran, an excellent instructor, and was on his way out of the navy to join the airlines. This was a big loss for the navy.

The T-28 Trojan is a big aircraft and is capable of performing carrier landings. I immersed myself in the smells of the engine's exhaust and the wild vibration that rumbled through the airframe at full power, along with the consuming environment of sweat, instruments and procedures.

Like most things, learning to fly in the navy requires a step-by-step approach. Initially we fly in the rear seat, literally under a bag attached to the canopy that blocks any view of the outside world. This is to simulate basic instrument flight. We learn to develop an instrument scan and to gauge the aircraft's performance through the interpretation of the various instruments. Next is the radio instruments phase, using ground based navigation transmitters to fly between known points or routes, still 'under the bag' in the back seat. Then we start flying in the front seat while our instructor is in the back. We are introduced into the world of visual flying and manoeuvres, landings and take-offs.

Before every flight, Ace and I discussed the management of emergency procedures. For me, this was where the foundation of Aviate, Navigate, Communicate was laid.

Ace would prepare me, lecture me and quiz me on this critical tool as we discussed the various failure scenarios. 'No matter what happens, Sully, you will always have time if you have altitude. The closer to the ground you are, the less time you have to manage the situation, so always try to gain some altitude as you assess your malfunction. Always be aware of your fuel state; if you have fuel then you have time, too. Even if the engine fails at low altitude, convert speed to height, wind your clock and assess your options.

'Aviate: maintain control of the aircraft first and foremost. Once the aircraft is stabilised you can Navigate: can I restart the engine, can I access a checklist, do I need to find an area for an emergency landing? Finally, once the other two are stabilised, you can Communicate: who you are, where you are, your situation and your intentions. This is the framework for the rest of your career, and I will constantly test you in establishing this discipline. Remember, anything can happen when you are flying solo, so I have to prepare you for that day because ole Ace won't always be there to save your ass!'

Being cleared to solo, to fly without an instructor, is a fledgling pilot's right of passage into the world of aviation. And flying solo in a high-performance navy trainer around the fields of Florida was certainly a big deal. As I progressed, Ace loaded me up more, forcing me to prioritise and to develop a 'coolness under pressure' approach to any situation. Routinely, he would simulate an engine failure by abruptly pulling the engine power back to idle; this tested my situational awareness, my response to failures, my memory recall of the emergency procedures, and my control of the aircraft's flight path as I recited the engine failure procedure: 'Fuel. Mixture. Mags. Simulate start.'

While I scanned the earth below for a suitable place to land my stricken Trojan, Ace's voice would inform me that the restart was unsuccessful, and I'd continue with the forced landing procedures. I would then nominate my proposed landing area, usually a crop field, and transmit my simulated mayday call over the intercom. Once all that was successfully completed and the Trojan was gliding towards an unsuspecting farmer's field, Ace would announce, 'Knock it off!' and I'd apply power and climb away in preparation for my next sequence of manoeuvres.

I worked hard and soloed on 14 October 1977.

Aviate. Navigate. Communicate. My first mantra – and it proved to be my saviour for a career in the air. It's the pilot's mantra from all over the world, and the basis for a fundamental discipline when operating an aircraft. Ace taught me well, and I salute him.

*

As we climb out from Singapore, I apply the mantra to our routine situation.

'Air, Engine Bleed 1 Fault,' Pete announces. He glances at me, one eyebrow raised.

This is when we laugh about me being a shit magnet.

To a layperson the fault might sound alarming, but the engine isn't really bleeding. This system syphons off, or bleeds, pressurised air from the engine and uses it to provide air-conditioning, cabin pressure and other system functions. The bleed fault indicates an issue with one of the valves connecting the engine to the system.

It's a simple procedure to reset the fault, and we proceed with the displayed steps to bring this system back on line. My job is to keep an eye on the aircraft's flight path while Pete actions the necessary steps. It doesn't take long.

Thunderstorms are forming on the horizon over the islands of Indonesia. Through the autopilot we modify our tracking to give them a wide berth as we fly south. It's easy to avoid these building monsters, and soon we're in clear air again.

After ninety minutes of duty time, I'm rostered to take my inflight break. I hand over control to Pete and exit the cockpit after Ross takes his position in my seat.

It's a long tour of duty for the pilots today, and I'm taking my share of required rest. Our rest area is outside the flight deck and is equipped with a foldable bunk, a fully reclining chair and a screen linked to the passenger entertainment system. Normally I'd visit the cabin before starting my rest, but the business-class meal service is in full swing, the carts blocking the aisles. I decide to check on my crew after my rest, and with that, I enter the cubicle and lock the door behind me.

I'm not really tired because I slept well the night before, so I select an episode of *Two and a Half Men* on the entertainment channel. Charlie has done something to himself that's causing him to scratch his groin incessantly. His brother, Alan, is disgusted but Charlie doesn't tell him the cause of his itch, only that he'll use any available implement to assist in scratching 'down there'. Later in the episode, Alan winds up in hospital, and Charlie takes the opportunity to expose his loins to the Indian or Pakistani doctor for a prognosis.

'Oh, my goodness! Your little jimmy is all crimson! Have you been applying anything to your genital region?' the doctor asks with wide-eyed incredulity.

Charlie smirks and replies he has only been applying actresses and waitresses to this region. He then admits to using hair dye to get the grey out of his pubic hair, but he leaves some silver bits around the sides to look 'distinguished'.

In a pontificating tone, the doctor replies, 'Mr Harper, we have a saying in my country. You can put a tuxedo on a goat, but it's still a goat!'

I'm laughing my ass off as the chime of the intercom alerts me to an incoming call from the flight deck.

It's Pete. 'Hey, Captain Kev, you're back on!'

'There in five.'

Still laughing, I exit the crew rest and peek through the curtains into business class. The cabin crew are busy clearing the last plates and coffee cups from the meal service. I won't disturb them with a visit. After a quick toilet stop, I request access to the cockpit to resume my duty.

The time is 1235 in Perth.

'A goat in a tuxedo is still a goat.'

I'll have to look for a good opportunity to use this line.

I'm relaxed and in a great mood. The flight deck is brightly lit from a cloudless sky as I enter. I'm at 37,000 feet and on top of the world.

# 3.

'Whatever you are, be a good one,' Abraham Lincoln once said.

I chose to be a pilot early in my life. For me, that contract was signed at age six. There was a Cold War and a Space Race, and the world was evolving and exploring and in conflict.

I was living in San Diego, California. My father was a chief warrant officer in the US Navy, happily married to my mother who lovingly raised five children and was a typical military wife in those times. I was the oldest of two boys, with two older sisters and a younger one rounding out our family.

On Armed Forces Day, my father took me and my brother to Naval Air Station North Island for the displays and open-house inspections of the warships – and, most importantly, the display by the Blue Angels. This flight demonstration team proudly showcases the precision and skill that all US Navy pilots are trained to perform. That day the Blue Angels were flying the F-11 Tiger: a supersonic, single-engine fighter that operated from aircraft carriers throughout the fleet. The Tigers were painted a high-gloss navy blue with gold lettering and numbering. They were the most amazing machines I'd seen in my young life – also the loudest and fastest. Watching them fly so close, so fast, so precisely … Well, the navy had me from that day. It looked like something a brave person would do, and I wanted to learn to be brave.

I wanted to be one of those pilots, to be ripping through the air, and I'd do anything and everything to be in one of those machines.

When I shared this dream with my father, he bluntly told me I'd only get there by pushing myself to excel. He had survived the Great Depression and was a navy veteran of World War II, and he'd worked hard to be a chief warrant officer.

I decided my path was set, and it was up to me to start that journey.

Soon I was a full-fledged aviation geek. At school, I was always near the top of my class in Maths and Science, and I grew close to a classmate who shared my ambition to fly in the military. Before I learned to drive, I would beg my parents to take me to the library so I could read issues of *Aviation Week & Space Technology*.

I came across a novel by Charles Nordhoff and James Norman Hall, *Falcons of France*. It was based on their experiences in the Lafayette Escadrille in World War I, flying Nieuports and Spads against the Hun over the trenches of No Man's Land. Naively, I concluded that I'd be a combat pilot and live that adventure.

The Vietnam War was in full swing. More books on modern combat emerged, such as *Thud Ridge* and *Flight of the Intruder*. Colonel Robin Olds, USAF Wing Commander, F-4 fighter pilot and MiG Killer, became a role model to me. His junior pilots would do anything for him – fly any mission no matter the risk – and he'd be leading from the front. He had a reputation for being at odds with his superiors, but he and his pilots were high achievers. I also read about MiG kills by navy pilots Duke Cunningham and Willy Driscoll over North Vietnam. Cunningham's famous quote, 'Train like you fight', would echo in the ears of many aspiring fighter pilots.

I built plastic aircraft models to do battle in my room, imagining what it would be like to fight the Red Baron, Adolf Galland or Colonel Toon. But I didn't really know if I could fly an aeroplane.

My chance came when I was on the high school tennis team in San Bernardino. We had a match in Perris, and a student's father offered to fly some of us down in his Mooney light aircraft. 'Anyone want to sit up front?' My hand came up first, so I sat in the front right seat on the way down. Despite my teammates' loud protests, I was allowed to fly the Mooney all the way back to San Bernardino as I explored its angle-of-bank limits. We'd lost the tennis match, but my flying was pretty good.

*

While in high school, I decided to pursue a degree in Aeronautical Engineering and to fly as a military officer. The US space programme was in full swing, and I marvelled at the Gemini and Apollo mission launches. The idea of flying further and higher dangled in front of my face like a big carrot; I wanted to become a test pilot or maybe even an astronaut.

To get there, I could follow one of three routes: attending a military service academy, getting a Reserve Officers' Training Corps (ROTC) Scholarship at a university sponsored by one of the services, or graduating from university then applying to become an officer candidate. I chose the ROTC route, setting my course to compete for a scholarship and attend the University of Colorado in Boulder.

I made it to the final phase of the selection process: an interview with a reservist officer. Despite all my preparation, it didn't go well – within minutes I sensed he didn't like me that much. Two weeks later the result arrived in the mail: I was a tentative finalist, meaning I had to wait for a finalist to turn down a scholarship. This was a huge blow. Never having experienced this type of failure before, I was demoralised and didn't know what to do.

I'd been accepted to the University of Colorado as an Aerospace Engineering student. My father tried to convince me to do a degree in pharmacy, which he was willing to fund if I quit ROTC and stayed in California. I told him I'd continue with my plan to attend the university of my choice and follow my dream, offering to cover food and board if he would pay for the first year's tuition. Reluctantly, he agreed. This was a significant amount of money for a modest middle-class, single-earner household with two children still in school. I would try again for an ROTC scholarship once I was enrolled.

It was a big gamble, but any other outcome was unimaginable to me. I had to believe in myself and follow my ambition, fighting to prevent the unthinkable.

At the tender age of seventeen, I didn't realise my 'plan' was akin to loading myself into a cannon. When I stepped off the bus in Boulder, I pulled the lanyard that launched me into the world of aviation – if I survived the cannon blast, I could be a pilot. I knew it would be hard but I welcomed the challenge. Alone, frightened and excited, I dragged a heavy suitcase towards my campus dormitory as I took my first steps towards my future.

The navy was expanding its nuclear power programme and was looking for 'hard science' majors to fill its ranks. At the admin office of the campus navy detachment, I enrolled in the navy ROTC unit in a non-scholarship position. I still had to comply with all the requirements as if I was on scholarship. As part of the enrolment process, I was asked, 'Are you interested in becoming a nuclear propulsion officer?'

'Of course!' I exclaimed with an eagerness that was perhaps a little disproportionate to my true interest in this career path. My field of study satisfied the navy's initial criteria. But the navy wouldn't consider granting me a scholarship until I passed the first year of my engineering course. I also had to attend Naval Science classes, wear my midshipman uniform to classes every Thursday, attend a marching drill every Thursday afternoon, and do physical training at 6 a.m. twice a week.

Our drills took place in fields next to the dorms, well within snowball range. I had to hide my smirks at the marijuana fumes that wafted over the marching midshipmen as our platoon leader's cadence competed with 'Light My Fire' blasting from speakers in the dorm-room windows.

The Vietnam War had just finished, so wearing a military uniform around campus singled me out for a few choice comments from students and some special treatment from a few professors. Being young, bulletproof and a bit blinded by ambition, I grew a thick skin and absorbed with humour the abuse for choosing to serve my country. I didn't think I fitted the mould of a 'baby killer'.

My illusion that I would match my high school academic performance was quickly shattered by the insane workload dumped on first-year Physics and Calculus students. I felt as if I'd been bound like an Egyptian mummy, taken a mile off the coast in a boat, thrown overboard and told to swim to shore in less than thirty minutes – my first priority to avoid drowning in homework, followed by getting it in on time. Fortunately, I had company: my roommate struggled with me, perched on the edge of academic extinction. My first-semester grade averages were just above the pass mark.

During the Christmas break, I travelled back to California. I knew what was waiting for me. Unimpressed with my academic performance, my father was primed to pull my funding, and rightly so.

We were having an emotional and frank discussion when I excused myself to don my midshipman's uniform, the double-breasted dress blues. I returned to face my father. The image of his son in the uniform he'd worn himself for so many years, imploring him for one more chance, was enough for him to give in. But even if he'd refused, I would have worked to fund my second semester. I was learning to be brave and not quit.

In late January I returned to campus desperate to succeed. To me, 'succeed' and 'survive' now had the same meaning. I was learning to be brave, and I willingly put my face into the academic meat-grinder.

As the semester drew to a close, I was invited to a formal interview with a navy instructor as part of the scholarship selection process. We discussed my ambitions versus 'the needs of the navy', and I told him I wanted to serve as a nuclear propulsion officer. Throughout the interview, the most important of my young life, I kept thinking to myself, *Don't mention flying. Don't screw this up!* 'Yes,' I said out loud, 'we must stop the threat of communism … Yes, we must confront the Soviet Union and the Eastern Bloc … Yes, I'm prepared to serve my country.' I had learned the lesson from my interview in high school. This time, I succeeded, but it all hinged on my grades.

During the summer break, I sweated on the daily arrival of the mailman. When my grades finally arrived, I ripped open the envelope and was elated to learn that I'd passed. My father expressed relief, but privately, he was concerned for his eldest son – my career path could easily place me in harm's way. He had experienced the dark side of war but never mentioned it to his family, and there was nothing he could say to me that would alter my ambitions. At least his financial burden would be reduced significantly thanks to Uncle Sam.

In a small ceremony at the start of my sophomore year, I stood at ramrod attention in my impeccable dress uniform. With my right hand raised, I swore allegiance to the United States. I was awarded a three-year scholarship and enlisted as a midshipman into the US Naval Reserve. My chest was bursting with pride. I'd made my plan work, but there was only a slight reduction in pressure to succeed – maybe 1 psi, in the nerd-speak of a second-year Engineering student.

*

Midshipmen completed six weeks of officer training every summer as a condition of the scholarship. The summer between my second and third year introduced us to the world of naval aviation. At Naval Air Station Corpus Christi, candidates were subjected to the most comprehensive physical and academic testing conceivable. Those who passed were selected to fly in the back seat of a training aircraft. My indoctrination flight was in a T-2C Buckeye: a twin-engine, basic jet trainer. This would be my first taste of military flying.

We were bussed to Naval Air Station Kingsville, Texas, and taken to the T-2 training squadron. The parachute riggers fitted us out in a flight suit, g-suit, helmet and oxygen mask. We then received a group briefing on what to expect during the flight and on safety procedures; we'd been trained in pressure chamber, ejection seat and water survival. Then we were bundled into the squadron ready room – a bunch of skinny twenty-year-olds in ill-fitting flight suits – to wait for our names to be called and meet our flight instructor.

As excited as we were, the overall emotion was worry. Would we puke under g-force during the aerobatic manoeuvring? Some had already returned ashen-faced, their suits adorned with the remnants of their last meal. I refused to eat anything until I completed my flight.

My name was called.

'Here, sir!' Time to fly.

I don't remember my instructor's name, just that it was summer and his flight suit was marked with perspiration. Out on the flight line, he led me to our jet and asked, 'Have you flown before?'

'Yessir, a light aircraft.'

'Did you get sick?'

'No, sir.'

'Have you got your barf bag?'

'Yessir.' I pointed to my g-suit pocket.

'Good. If you feel queasy, go on cold mic and make sure you aim well. You get to clean out the cockpit if anything happens!'

I don't blame him for not wanting to listen to me lose my lunch.

The communications system on the T-2 is like that on any military aircraft. You can talk freely into your oxygen-mask, on cold mic, by pushing a button, or you can activate the microphone by flicking the hot mic switch and you will hear everything that comes out of each microphone.

My instructor introduced me to my 'guppy', the nickname for the T-2. It was painted orange and white, the colours of the Naval Air Training Command, with a large white B on the tail. I gently slid my hands along the wing's leading edge, then relished every step as I climbed the hand-and-footholds up the side of the jet and immersed myself in the back seat. I was followed by our enlisted plane captain who connected my oxygen mask, communication cord and g-suit as I waited impatiently for my instructor to return. (A plane captain is a member of the ground staff assigned to each departing aircraft.)

After inspecting the exterior of our jet, my instructor showed me the communications panel and microphone switch, then prepared my ejection seat by removing the safety pins. With a questioning look, he flashed me a thumbs up, which I returned. He was quickly off to strap into his seat.

Soon we were connected on hot mic. In the hypnotic monotone of a seasoned navy instructor pilot, he recited the checklists from memory. I listened intently but was also mesmerised with my environment: the instrument panel loaded with engine performance gauges, the large blue-and-black attitude indicator, the feel of the control stick and throttles under my fingertips.

My instructor announced, 'Okay, we are cleared to start.'

Signalling with his fingers, the instructor started the engines in harmony with our plane captain. The turbines rotated as the turbojets came to life, and I welcomed their high-pitched whine through my helmet earpads. I deselected the hot mic switch and cracked the oxygen mask from my face, which was dripping in sweat. Inhaling deeply, I sucked the high-octane jet fumes into my lungs; into my being. I would never tire of smelling the exhaust from any of the jets I flew. At that moment, I knew this was my destiny.

Once the instructor had completed the after-start procedures, he pointed his thumbs opposite each other, signalling the plane captain to remove chocks – the rubber blocks holding our wheels stationary.

Like the Blue Angels departing to start their display, we taxied forwards under the direction of our plane captain. As we passed him, he snapped a salute. My instructor snapped one back, and I did too – it was my first opportunity to perform this ceremonial salute, so I wasn't going to let it pass. The young plane captain smiled and walked towards the hangar, shaking his head at the nerd midshipman in the back seat of his T-2 and praying he wouldn't have to clean it later.

We taxied to the runway on cold mic with the canopy up, providing some welcome relief from the searing heat and humidity of southern Texas. The T-2's air-conditioner wouldn't be effective until we were airborne and ram air was forced down its inlet. I was drenched in sweat under my flame-resistant Nomex flight suit and gloves.

'Clear canopy,' came the command over the intercom.

I synched down my oxygen mask and selected hot mic. 'Clear!'

As our Buckeye entered the runway, we were cleared for take-off. We stopped briefly on the centreline while the engines went to full power, then the brakes were released and we started to roll, accelerating to lift-off speed.

We were airborne. The landing-gear handle came up with a *thunk* and the flaps were retracted as we accelerated, climbing away from the base.

'How are you feeling?' came over the intercom, my instructor's visored helmet looking at me from one of his mirrors.

'Great,' I murmured and flashed a thumbs-up. I hadn't experienced sex yet, but this had to be better!

The plane levelled off, keeping low, and the airspeed increased towards 300 knots. A broken layer of cloud was above us, and I figured our level-off was to stay clear of it.

Wrong – when we reached 300 knots, my instructor pulled the jet up steeply and punched through a hole in the cloud base. As the clouds streaked by, I looked over at the condensed moisture obscuring our wings from the g-force and pressure change over the wing.

My instructor performed a wingover, an aerobatic manoeuvre in which a steep climb is followed by a vertical flat-turn and change in direction. The nose of the jet was 45 degrees above the horizon as he pulled up into the wingover then released the back stick and g, rolled the wings to 90 degrees of bank and I watched as the nose sliced back down towards the horizon.

It's my first opportunity to experience g. A quick glance at the g-meter showed a reading of +4. I felt the heavy weight push me hard into the seat as my body suddenly weighed four times more than normal and I loved it.

My instructor couldn't see the cheesy grin beneath my sweaty oxygen mask. Again he glanced back, and I flashed another thumbs-up to let him know I was handling the g-force and his aggressive manoeuvring.

'Okay, you've got it,' came over the intercom, as the control stick wiggled.

'I've got it.' I returned the wiggle. *Oh yeah … I've got it.*

For the next half-hour he talked me through the instrumentation and how to fly with the wings level. He demonstrated more aerobatic manoeuvres while observing me through his mirrors; I was then allowed to try the manoeuvres while he assessed my ability to maintain three-dimensional spatial awareness.

I was in my element. It may sound like a joy ride but it was far from that. I was being assessed on my aptitude to fly and for consideration to request flight training after I graduated.

Despite my elation, I knew that this flight was a critical piece in my future career.

All too soon, we descended to the training field. As we returned to the hangar with sweat stains on both our flight suits, my instructor looked at me. 'How was that?'

My enormous smile was my answer, followed by a respectful, 'Thank you for a great hop, sir.'

He smiled back and gave a few nods as if to say, *You'll do.*

In my navy career, I would accumulate a total of 500 fight hours on the T-2 as a student pilot and then as an instructor.

At the time, the brass ring on my merry-go-round was in sight – I just had to graduate before I could grab it. I started my junior year with another full plate of subjects, though our professors treated us a bit better now we'd survived two years of hell and were starting to specialise.

That year, I participated in the NROTC Aviation Indoctrination Program: forty hours of private flying at a local airfield that would help me obtain my private pilot's licence. This meant getting up at 4 a.m. in the Colorado autumn and winter, then driving to the local airport where I launched at sunrise:

my first 'dawn patrols', but certainly not my last. On touchdown, I'd quickly debrief with my instructor, jump into my borrowed car and blast back to campus for my first classes by 8 a.m.

I passed my licence-check ride the following year. My first flight as a private pilot was to visit my fraternity house, accompanied by my roommate as 'copilot' and bombardier, and display my aeronautical skills to my frat brothers … enough said.

In May 1977, I completed my Bachelor of Science and was commissioned as an ensign in the US Navy. My mother and father witnessed their proud son embark on his aviation career.

# 4.

On Qantas Flight 72, conditions are perfect; the air mass is smooth, cloudless and sapphire blue. We're still over the Indian Ocean, but the red shoreline of Western Australia's northern coast is now clearly in view.

I first visited WA in August 1981. I was on my first operational cruise on the USS *America*, which involved a port visit to Perth.

The aircraft carrier signalled its impending arrival with a sixteen-aircraft formation. I was in the trail formation of four F-14s, and the other three formations joined to form a 'diamond of diamonds'. The cloud base was relatively low, making our procession even more spectacular as this massive formation flew around the city at an altitude less than 3000 feet. Traffic jammed everywhere as drivers stopped to watch us fly past. I couldn't look down as I was flying in the 'slot', maintaining my position relative to the other three F-14s as we circled the city. The next day, the *America* set anchor near Fremantle and our port visit began.

Life aboard an aircraft carrier is a spartan existence. For starters, you have more than 5000 roommates – and, in my time, they were all guys. There was no privacy, no creature comforts. The shower and toilet facilities were communal, and sometimes the ships' plumbing got blocked and those areas were flooded with backed-up sewage. The ship operated around the clock, whether or not the aircraft were flying.

The ceiling of my small two-man stateroom was connected to the flight deck. This meant my precious few hours of sleep were disturbed by sounds of aircraft taxiing, catapults firing, arresting gear trapping landing jets, the muffled roar of engines at full power, and the scraping of tie-down chains as the blue shirts marched along to secure another jet to the flight deck .

The food in the officers' mess was below average. On my first cruise, I survived on celery and peanut butter; on the second, 'sliders'. This is navy code for a cheeseburger, and they were made available later in the evening, well after the normal dinner service was complete. For junior officers, our days revolved around having a slider as often as our flying schedule allowed. If the soft-serve ice cream machine ever broke, there was a mini mutiny in the officer's mess.

Before our port visit in Perth, I'd known very little about Australia. In my elementary school geography class, Australia was part of Oceania, on the other side of the planet. Later, I read history books that depicted Australian soldiers as fierce warriors, both during the trench warfare of World War I and as the Rats of Tobruk in World War II. Warner Bros. cartoons portrayed Australia as a wild place with boxing kangaroos, Tasmanian devils and boomerangs. Bugs Bunny had many encounters with the tornado-like Taz, one of my favourite characters.

I found myself impressed with Perth and the people of Australia. The pace of life was slower, the climate was clean and the beer was cold. My duties aboard the ship as a junior officer precluded me from spending much time ashore. It took a while to look the correct way when crossing the busy streets and there were some close calls. I had enough time to enjoy a cold Swan Lager or two, but the high alcohol content was more than our beer-free bodies could manage (our ships were 'dry'). I hoped to return someday to enjoy a summer.

*

Nearly a year before my journey to Australia, in October 1980, my transition to the F-14A Tomcat at the Replacement Air Group squadron, VF-124, at NAS Miramar – otherwise known as 'Fightertown' – in San Diego, California, was nearing completion. This would be my final step before I joined an operational squadron and carrier deployment.

The Western world was still embroiled in the Cold War, so I had a good chance of being involved in armed conflict somewhere on the planet. The tempo of training was intense – and proportionally dangerous to cater for this level of political uncertainty. As always, the rapid power-projection afforded by a carrier taskforce meant that the navy and marines would be the first on scene at any potential hotspot. At the age of twenty-five, I was positioning myself at the pointy end of the spear of American foreign policy.

The F-14 course would prepare me for fleet operations. Part of the syllabus included night carrier qualification and air combat manoeuvring. The navy had experienced an increase in accidents caused by pilots losing control through aggressive manoeuvring as a result of increased air combat training. Although mandatory spin training was introduced in basic jet training, it wasn't enough to cater for the various scenarios a fighter pilot could encounter when pushing his aircraft to its limits during aerial combat.

The Navy Test Pilot School developed a training course to address out-of-control flight, and the aircraft chosen for this was the T-2 Buckeye. The T-2 was a twin-engine jet trainer and had proven its ability to enter and recover from upright and inverted spins. I'd flown it during basic jet training as a student, and later became a T-2 instructor pilot. This course in out-of-control flight became mandatory training for all pilots transitioning to fighters.

No fighter pilot was worth their salt if they hadn't taken their aircraft to the limits of its performance and beyond. The pilot had the ability to exceed these limits if required, but they needed a skill set to regain control if it was lost. This innovative programme ensured that pilots could recognise and recover from upright and inverted spins, accelerated and yaw-induced departures, and zero-airspeed departures – extreme manoeuvres that we might encounter during air-to-air combat.

My instructor explained the signals the aircraft would give as it approached its critical angle of attack: the point where the airflow over the wing separates and the lifting force of the wing is lost. The T-2 was equipped with a rudder pedal shaker system, another physical cue to the pilot that the boundary of uncontrolled flight was approaching.

The rudder pedal shaker was linked to the angle of attack system (AoA). This gauge measures the angle of the airflow over the wing. Every wing has a critical angle of attack where the airflow separates from the wing and the aircraft stalls; basically, it starts falling like a bird that had lost its feathers.

The manoeuvres required me to fly the aircraft past its limit, forcing entry into the out-of-control phase, then to recover using the F-14's spin procedure. As the pilot, I had full feedback and control through the traditional cable and pulley system that connected my control stick directly to the aircraft's wing and tail. Tactical aircraft are also fitted with a rocket-powered ejection seat, a pilot's 'get out of jail free card', which we wouldn't hesitate to activate if we were still out of control and below 10,000 feet.

Over the eastern desert of California, I routinely pushed my aircraft into extreme territory, exceeding the critical AoA. I spun and tumbled the T-2 beyond its operating limits under the supervision of my specialised instructor pilot. The stuttering rudder pedals hammered at my feet as I pushed the T-2 past its limits of controlled flight and into the departure and spin phases. The gyrations of the plane were wild and similar to those of a Pitts Special at an airshow. It was physically challenging to maintain stick and rudder pressure as the jet tumbled, and I waited impatiently for the instructor to command, 'Recover!'

My athleticism was tested as I explored the positive and negative g limits of the T-2, and I spent a lot of time hanging in my straps after the negative g accelerations. I knew that if I was to be a successful fighter pilot, I would have to master the recognition and recovery skills introduced in this training. It was an essential part of my progression and essential for my future selection to attend Top Gun.

At the completion of this short course, I was more confident that I could survive most of the adverse flight regimes I might encounter during the extreme manoeuvring that lay ahead in my transition to the F-14. This survival skill set became second nature during my military flying, but I'd never needed to use it in the mundane world of commercial aviation.

Until today.

*

In the cockpit I give Ross a friendly punch on the shoulder and jerk my thumb back, my signal for him to give me back the captain's seat. I slide in then secure the lap and crotch belts of my safety harness. I leave the shoulder straps off as I listen to Pete's handover briefing to me. As he talks, his fingers dance over the keys of his navigation computer. 'Weather is fine at destination and at our divert fields. Progress to Perth is showing an on-time arrival. We're maintaining the increased cruise speed to make up time, and we climbed to 37,000 feet while you were on your break.'

My office is serene, just the way I like it. Commercial flying is all about flying safely from A to B. Boring is good in this job.

On my own computer I follow Pete through the briefing. Ross has excused himself for a toilet break. He'll be on duty with me for the next hour and twenty minutes, and Pete will start his rest period after this briefing.

Pete points to the left of the aircraft's nose to highlight Learmonth, our closest airport, near the town of Exmouth, Western Australia. I scan my navigation display, confirming the symbology for the airport and its position relative to our flight track.

I nod and give my first officer a thumbs-up. 'Have a good rest, Pete. I'll call you up thirty minutes before our descent point. I have control.'

Pete exits the flight deck, and I'm alone.

The time is 1239 in Perth.

I scan my instruments and the systems panels above my head; all the lights are out, indicating the systems are functioning as designed. A single chime alerts me to a call from Ross outside the cockpit, requesting access. After confirming his identity, I remotely unlock the cockpit door. As he slides into the right-hand control seat, I nonchalantly ask him if he took the time to wash his hands. He chuckles as he straps in and leaves me guessing.

'No change, Ross,' I report, meaning there have been no modifications to the aircraft parameters that existed before his toilet visit.

*De-deet. De-deet. De-deet*, shrieks from the cockpit speakers, signalling a warning that the autopilot has disconnected. The red master warning light is illuminated.

The time is 1240.

'Shit,' I mutter with disdain. I must fly the aircraft manually now using my sidestick, trying to maintain my altitude while searching for information to explain the autopilot fault. I scan the instruments to find the cause as I direct Ross to cancel the red warning light and accompanying aural warning.

I call for the reconfiguration procedure to address the autopilot problem while instinctively engaging the second autopilot. I need to unload myself from the task of maintaining altitude and devote my attention to troubleshooting the fault messages appearing on our computer screens. Our altitude deviates slightly during my manual flight but quickly stabilises with the autopilot corrections. My instrument scan doesn't provide the reasons for this malfunction.

Before we can action the other faults, the relative calm of the flight deck is shattered:

*Ding-ding-ding-ding-ding!*

*STALL! STALL!*

*Ding-ding-ding-ding-ding!*

I'm thinking, *Now what?*

Perplexed, but still calm, I scan my instruments to find the causes of these serious warnings generated by the aircraft's computer systems. The red master warning light is on again, and a red 'OVERSPEED' message is emblazoned on top of my primary display. The Overspeed warning alerts pilots that the high speed limit for the aircraft has been exceeded.

The altitude and speed indications on my screen are jumping up and down; the computers can't display the altitude and speed that were normal seconds before.

Now my navigation display has gone blank. Critical systems are shutting down, and the computers aren't telling me the reasons. The loud stall warning indicates that we're at a critically slow speed while the overspeed indicates we're at our maximum speed too.

It makes no sense to me. Moments before, we were in stable flight; nothing in the outside environment has changed.

We're lucky that this is a daylight flight. I can see the horizon; one minute ago I was in straight and level flight and my cruising speed and power were stable. We're in a calm and clear air mass with no turbulence or cloud. Are my computers crashing or reacting to something I can't see or control?

The time is 1241.

'Ross,' I say, 'call the forward galley.'

I ask Pete to return to the cockpit – we need all hands on deck to manage this confusing mess.

I take a deep breath and exhale slowly as I keep scanning my instrumentation in the midst of the continuous stall and overspeed aural and visual warnings. I'm searching for clues to indicate the cause of our worsening situation. I'm still calm and in control, but my confusion is building.

*

In the cabin, Pete has started his rest break with a visit to the forward galley, chatting with Lisa, our customer service manager, and drinking a coffee. The meal service has finished, and the remainder of the cabin crew are starting their rest periods before our arrival into Perth.

In the rear galley, Fuzzy Maiava is cooking his crew meal in the oven. Two off-duty staff members, Diana Casey and her husband Peter, have joined him, and they're having a chat while sipping on glasses of wine.

Diana is an active customer service manager in the domestic division, and Peter is a senior captain on the Boeing 737 fleet. Both are based in Brisbane. Travelling with them are their two teenaged daughters, seated in another part of the cabin, and together they've been on a holiday to celebrate Peter's successful recovery from a long-term medical issue.

Adding to this festive mood, Fuzzy lets the Caseys know that today is his birthday and he's planning to have some celebration drinks in Singapore on our return.

Fuzzy notes the timer on the oven: thirteen seconds until his meal is ready.

Throughout the full cabin, passengers are moving around, visiting friends in other parts of the plane and waiting patiently to use the toilets. Small children are unbuckled, playing on their parents' laps. There's a lively, party atmosphere – it's the way air travel should be.

*

While Ross is contacting Lisa through the interphone system, I disconnect the autopilot and revert to manual control. My speed and altitude tapes are still jumping around, so I can't use them with any accuracy. I shift my scan to a group of secondary instruments for another source of speed and altitude information, and I crosscheck them with what's displayed on the first officer's side. They match, meaning I can use them safely while I control the aircraft manually with my sidestick.

I've figured out I'm dealing with an abnormal condition that Airbus has labelled 'unreliable speed'.

Unreliable speed means that the information values displayed on my main airspeed and altitude instruments are not reliable or that they are false. This is not displayed as a warning or caution message because the automation doesn't make this critical assessment; it is up to the pilot to interpret. In the past, this condition has usually resulted from something blocking the airspeed-measuring probes mounted on the fuselage near the aircraft's nose. The non-normal condition requires me to fly the aircraft manually, without the autopilot and auto thrust, while we try to address the problem.

My concentration is elevated now as my fingers lightly touch the sidestick while I manipulate the controls to maintain my altitude. We rarely fly the aircraft manually at these high altitudes, and my stick inputs need to be small and smooth. The incessant warnings keep blaring in the flight deck, no matter how much we try to silence them.

I'm getting mildly annoyed, but I must focus on controlling the aircraft in this loud and distracting environment. Confusion is creeping into my brain as I scan my instrumentation, but the computers are much more confused than I am – to the point that they believe we're in an extreme danger and won't stop warning us.

I assess that I'm surely not in danger from anything outside the plane. I don't agree with the computers that the aircraft is in a threatened position.

The time is 1242.

My concern is growing. These are serious warnings that by design require immediate action, but looking out the forward cockpit window confirms we aren't under threat. These warnings *have* to be false. The cause isn't displayed and I am dealing with the results of their confused state. I won't act until the computers identify the malfunction, as we were trained to do in our Airbus course.

The cockpit environment is loud and distracting. I can't access the information that the plane's computers are generating and using. Somehow, though, the computers are reacting to information that's valid enough for them to disconnect the autopilot and issue all these warnings.

We haven't come across an event like this in our Airbus training. It seems the many computer systems that control the aircraft are shutting down, but we don't know why. The highly automated systems, designed to make our job of flying the aircraft easier, are failing.

I must maintain safe control of the aircraft as I've been trained to do, but I know I'm being thrown into uncharted territory.

# 5.

The time is 1242:27.

I'm thinking about the next step in the checklist procedure for unreliable speed, removing the yellow flight director bars and disconnecting the auto thrust. At the moment, I'm managing to maintain my altitude by using the secondary instruments and the first officer's display, but it's like steering your car from the passenger seat. I am hoping this is a transient hiccup in our automation and I can fly like this if it doesn't fix itself.

I try to make sense of what is happening, methodically and clinically.

Ross is talking on the interphone; he's made contact with the cabin to coordinate Pete's return.

Suddenly, the nose of the plane dips slightly. Instinctively, my grip on the sidestick tightens and I'm on high alert; I haven't commanded that movement.

*

Pete is still in the forward galley finishing his coffee. He stiffens, sensing an abnormal vibration through the floor as if the tail is fluttering. The cabin interphone rings; it's Ross calling to get Pete back to the flight deck. Lisa answers the interphone call, just as Pete perceives a change in the engine's pitch.

The group of off-duty Qantas employees in the rear galley also hear the engines' pitch change – a decrease in sound that has, in their experience, signalled the commencement of descent from cruising altitude before we reach our destination. They look at each other and at their watches, and they're perplexed; there are still two hours of flight time before our arrival into Perth.

*

In the seat of my pants, I feel another uncommanded movement. 'Don't you …' I growl at the Airbus.

But I don't get to finish my stern warning to the aircraft's computers. The plane's nose begins to pitch down, violently and aggressively, accelerating its rate of movement like a roller-coaster about to speed down its tracks after reaching the top of a loop. The negative g-force is increasing and elevating me out of my seat.

In the right seat, Ross struggles to understand what's happening as the plane accelerates its nose-down trajectory. Then, spookily, the contents of the flight library beside his seat rocket towards the deck's cramped ceiling.

The leather- and plastic-covered manuals are heavy, but the forces generated by the aircraft's violent manoeuvre have made them weightless – Ross has to swat them away like a persistent swarm of flies. Then they're thrown to the rear of the cockpit.

*Thud.*

They explode on impact with the floor, scattering their contents behind our seats.

I grunt, my jaw clenched, as I yank my control stick to the full limit to stop the pitch-down. Because my shoulder harnesses aren't fastened, I'm bracing myself on the instrument panel glareshield to keep myself seated. I have to push hard to maintain my seated position.

'Shit!' Ross hisses as the plane's nose carves through the horizon.

There's a roar as the Airbus' huge wing moves through the air as it pitches down. Then a crash in the cabin reverberates through our reinforced cockpit door.

*

Pete hears an enormous *bang* as he and Lisa are elevated off the floor. At first they're surprised – what's happening? Their surprise turns to shock as they're slammed into the ceiling.

The highest g-forces are generated in the rear galley. Fuzzy and the Caseys are propelled into the ceiling with a greater impact than Pete and Lisa. The three are stunned by the impact; they remember going up but not coming down. They've been knocked unconscious.

Throughout the cabin, the scene is the same. It looks like NASA weightless training for more than a hundred astronauts.

But this aeroplane is manoeuvring violently, with massive g-forces, and the cabin walls aren't cushioned. People are screaming. Luggage from the overhead lockers crashes onto some passengers' heads, while other passengers are hurled towards the ceiling.

Abruptly, the screaming stops. It sounds like a horror movie where a person is attacked by an alien, screaming in terror until the violent creature ends their life. Those unrestrained by seatbelts are pinned helplessly to the ceiling.

The cabin interphone in the front galley, still on an active call to the flight deck, records the mayhem on the aircraft's voice recorder black box.

*

I'm not feeling calm and in control anymore. The plane is out of control; something is terribly wrong.

My aggressive use of the sidestick hasn't stopped the nose from continuing its downward plunge. I know that people in the cabin must be trapped against the ceiling. As I brace myself in my seat, my windscreen is filled with the blue of the Indian Ocean. A terrible glimpse of a possible outcome is projected into my consciousness – am I going to wind up in the water? Is my life going to end today?

It seems like time is standing still. I'm presented with a choice: keep pulling back on the control stick, even though it isn't generating any response, or let go.

Instinctively, I release my pressure on the sidestick. Out of my subconscious, a survival technique from a previous life rises up: *Neutralise!* I'm not in control so I must neutralise controls, return the sidestick to the centre and not make any input. I never imagined I'd use this part of my military experience in a commercial airliner.

An eternity passes. The plane is on a collision course with the Indian Ocean, and my sidestick seems as useless as a piece of overcooked spaghetti. Is my control stick completely dead? Gingerly, I probe for control with small sidestick inputs and hope like hell I'm granted control again, while the deepest part of my brain records this moment: the potential end of my life.

Finally, I feel some response.

The computers return control to me as I move my sidestick. I don't want to move too abruptly, fearful of hurting the passengers and crew, so I move as slowly as possible to bring the nose up. Our nose attitude is almost 10 degrees down. The aircraft has lost 690 feet of altitude in our dive as I carefully level out. It's the most gentle manoeuvre of my flying career. My hope is to smoothly lower people from the ceiling.

'What the *fuck* was that?!' Ross exclaims as we begin our crippled climb back to our cruise altitude of 37,000 feet.

A simple yet profound question, and there's only one possible answer.

'It's the PRIM,' I reply, using the abbreviation for the primary flight computers.

The pilot must interact with the plane through the computers and it is the PRIMs that move the control surfaces. They have decided to move them aggressively, despite my command to stop them.

I know these computers initiated all this mayhem, because what else could it be? I certainly didn't command it, and the outside environmental conditions didn't cause the plane to dive. But although I've lamely answered Ross's question, my response doesn't do enough to explain what's happening to us.

'Handing over, Ross. I'm putting on my shoulder harness.'

He takes control of the plane as I urgently reach back, grab my harness and push the metal clips into the buckle at my waist. I also pull my lap and crotch straps tighter. Because I don't know what's causing the aircraft to go crazy, I must be secure in case it happens again.

For a few seconds, Ross gets the opportunity to manually fly the A330 at high altitude. Not many commercial pilots can experience that these days, especially after a jet upset.

'I have control,' I say as I take over again from Ross, allowing him to secure his shoulder harness.

I look over at Ross. His eyes are flared wide as he scans his instruments, searching for clues. His gaze is locked in tunnel vision, as is mine – eyes open but maybe not seeing.

I feel like I've just skulled 100 double espressos. My body feels electrified, my eyes are jittery and vibrating, and my skin is hot.

The flight deck is a mess and so are our bodies, charged with the fight-or-flight chemicals released during our violent, life-threatening dive.

'Take some deep breaths, Ross,' I coax him. 'Hold it. Exhale. Keep breathing.'

We need to calm down, to try and overcome our nervous system reactions so we can function and think.

'Are you okay?' I ask. 'Seatbelt signs on and make an announcement, Ross.'

Somehow, Ross has beaten me to the punch – during the pitch-down he reached up and swiped the fasten seatbelt switch. He announces over the PA system, 'All passengers be seated and fasten seatbelts immediately!'

I still don't know why the computers took this shocking action. My eyes keep rapidly scanning the instruments, but the Airbus continues to withhold its secrets.

I accept that we're in danger and try to keep a cool head. We've survived a violent loss-of-control, and we are in the dark. As the captain, I must try to return everything to normal and make sense of our deteriorating situation. The aircraft systems designed to protect us are failing and trying to harm us. Why?

*

Although I had tried my best to be gentle, the many passengers pinned to the ceiling came crashing down as positive g-force was restored by the aircraft's recovery manoeuvre. Throughout the cabin, people are screaming and moaning.

In the front galley, Pete and Lisa also crash to the floor. Pete comes down first on his face and shoulders, then Lisa lands on top of him, shaken but unharmed.

Stunned, his eyes watering in pain, Pete gets up onto his hands and knees. It looks like a bomb has exploded in the forward galley. The steel locks on the meal carts' doors are torn open, disgorging their food trays and garbage onto the floor.

Likewise, bottles of wine have shattered, and red liquid covers the floor and walls like pools of blood. The air-conditioning ducts are exposed in the ceiling, their plastic cover panels strewn throughout the cabin.

There's even more devastation in the rear galley located near the aircraft's tail, which bore the full brunt of the massive force generated by the dive. The Caseys and Fuzzy are lying on the galley floor, badly injured. All three remember being elevated by the negative g-force, but none can remember coming down. Diana's shoulder is burning in pain. Her husband lies next to her, bleeding from a large gash on his forehead. She hears moaning close by – Fuzzy is groggy and in agony. Diana is horrified to see part of his leg bone protruding through his torn pant leg.

Peter turns his head to meet his wife's eyes. 'Are you alright?'

She nods, and they help each other sit up before propping Fuzzy up, too. Dazed, they slouch against the stowed metal meal carts as much as their injuries allow.

Peter, an experienced commercial pilot, ponders their position. What just happened? The seatbelt sign wasn't on, and the pilots didn't issue a turbulence warning. He concludes that the plane has experienced an unforeseen, un-forecast turbulence event. What else could generate this much movement and force?

*

The time is 1243.

Ross and I are still in the game; we have to be. Although our bodies remain electrified from the fight-or-flight response, we must get down to the business at hand. Our computer screens are lighting up with amber-coloured fault alerts. We're 154 kilometres to the west of Learmonth Airport.

I take a deep breath and direct Ross to begin our reconfiguration procedures in order to deal with the failures we're starting to see on our screens.

Methodically, we address the alert messages displayed on our screen. There are faults with the navigation portion of ADIRU 1, PRIM 3 and the control function of PRIM 1.

But it's difficult to implement the procedures. I'm not allowed to use the autopilot as a result of the 'unreliable speed' condition. Because I am flying manually, my workload must be split between monitoring Ross and maintaining a safe flight path. Likewise, Ross must watch what I am doing as he runs through the reconfiguration. Meanwhile, we're both distracted by the continuous warnings blaring out of our speakers. This routine is made even more complex by the uncertainty caused by our failing computer systems, as we try to solve a puzzle with missing pieces.

Our cockpit environment is degrading. Words can't describe the workload we're faced with. There are more problems displayed with the autopilot, a malfunction in one of the computers responsible for providing position information, and a problem – as I'd expected – with one of the three PRIM computers. Methodically, Ross and I perform what we hope will be life-saving surgery to our critically ill aircraft systems.

I'm slowly climbing back to 37,000 feet while using the information on the first officer's flight display to confirm my speed and altitude. My airspeed and altitude information remains erratic. All the while, I listen carefully to Ross as he works through the error messages. In vain, I'm trying to use my systems knowledge to connect the dots and get things back to normal. But even though I'm still baffled, I don't let my concentration wander from the immediate tasks before me. I don't have time to worry, to feel frightened or to contemplate death, but I must suppress a bubbling anger at the overwhelming position I have been thrown into and without warning.

Ross reaches a critical point for the next failed system. We are now required to interact with the third primary computer, PRIM 3. The PRIMs move the control surfaces via hydraulic fluid lines, similar to your car's power-steering system. Resetting the PRIM will move the valves that control these dedicated hydraulic lines as well as reset the system logic of the flight control system.

Ross needs to reset the control switch for PRIM 3. To do this, he must consult a special paper checklist. He's searching the case at his side that used to contain his paper manuals – but they're all behind him now, strewn over the cockpit floor.

I can see him in my peripheral vision, searching for his checklist, and I silently grab mine for his use. Even though the PRIM reset is simply a matter of recycling the switch off then on, the checklist procedure is full of cautions and advice. I have some previous experience in performing this reset, and we've practised it in the simulator during our training.

Ross is focused as he finds the relevant section for the reset procedure. He reads it aloud first, and I nod in understanding. He reaches up to the overhead panel, finds the switch marked

'PRIM 3', then looks at me to confirm he has the right one, and I nod again.

This switch is special, covered by a small guard that prevents inadvertent contact. Ross carefully raises the guard and pushes the switch to 'off'. There's a noticeable *clunk* as the computers adjust to the reset.

Immediately, the stall warnings stop, and I make a conscious note of this link from them to the PRIM 3 system.

For a few precious seconds, the flight deck is silent. Ross pushes the switch back in and the reset is complete. We're also back at our cruise altitude.

I reach down to access the lower computer screen in the centre panel, where I select an overview of the flight-control system. All the various aircraft systems can be seen here, and green lines and boxes indicate normal operation. After the reset, all the components of our flight-control systems are showing 'green' – nothing abnormal has registered here.

In the time it takes for me to deselect the display, the warnings start again.

*Ding-Ding-Ding-Ding-Ding! STALL! STALL!*

# 6.

In the rear galley, Fuzzy and the Caseys realise they're still in danger. The engines are changing pitch again; they must secure themselves as best they can while seated on the floor.

In the forward galley, Pete rises up on his knees as the floor starts to vibrate. The airframe is fluttering as it did before the first dive. Pete heroically takes action to protect Lisa. He reaches out to grab her tightly around her waist with one hand; with the other, he pushes up against the galley rail and wedges their bodies beneath it.

People all around Pete and Lisa begin to scream again. They've also realised this vibration signals the start of another dive.

Determined, Pete braces and waits.

*

The time is 1245.

The aircraft's nose is starting to move. Instinctively, my grip on the sidestick tightens. 'Don't you do it!' I growl under my breath.

The computers aren't listening to me.

For the second time, the nose pitches down, propelling us towards the water below. Once again, I respond with full back stick as Ross and I are pushed towards the ceiling. Our shoulder harnesses help this time, but I must still brace against the instrument panel with my right hand to stay seated.

I know now my input is a useless gesture; the computers won't grant me my command to stop the nose from pitching down. They're in control, and they're *very* confused.

'You fucking piece of shit,' I mutter at the computers as I'm once again confronted with a loss-of-control situation. I neutralise my sidestick as in the previous pitch-down – and once again, I get a windscreen full of the Indian Ocean.

I'm really pissed off now. In a flash, I suppress a growing fury as a harsh reality dawns on me: I'm in a knife fight with this aeroplane, and it has cut me twice. Is it really trying to kill me? I feel like I'm fighting with both hands tied behind my back, but I will not go down quietly. With a focused determination fuelled by my body's chemical startle response, I vow to fight to the death if that is my fate.

*

The horror of the second pitch-down is thrust upon the 315 passengers and crew. They're like a terrified child experiencing their first bumper-car ride at a theme park, waiting for the next collision and dreading it.

The seatbelt sign has been on for a few minutes now, and those who were able to have scrambled back to their seats and secured themselves. Those who are conscious have braced for the second of their sickening rides; others are severely injured or unconscious, lying helplessly on the floor. The conscious passengers struggle to hold them down, reaching out as far as their seatbelts allow, while the negative gravity elevates every unsecured person and object towards the ceiling. Many are screaming as the aircraft lurches down again, accelerating.

Will this be the one that ends their lives?

Still in the front galley, Pete is braced to fight the forces that lifted him off the floor a few minutes ago. As the floor presses into his knees, he maintains his tight grip on Lisa's waist with one arm, while keeping a firm hold on the stainless steel railing. Together, they elevate off the floor.

Pete holds on courageously through the haze of *what-the-fuck-is-happening*. He's aware that the two pilots are busy fighting for the lives of everyone on board, and knows he should be on the flight deck to assist. He'll go there as soon as he can. For now, his duty is to protect Lisa and survive this second dose of violence.

As the second dive commences, Fuzzy and the Caseys are sitting on the floor, their backs pressed to the sides of the cart stowage area under the galley benches. Unsecured, they're again subjected to the full gambit of lifting force – they rise up into the air before they're thrown back down. Stunned again after ricocheting off the ceiling, they lie like three ragdolls on the debris-covered floor as if carelessly tossed there by a bored child.

Peter Casey pushes himself upright, his vision blurred by the blood flowing into his eyes. Through the searing pain of his injuries, he realises that this isn't a turbulence event – something more serious is happening. Deeply worried, he knows that he, Diana and Fuzzy must quickly secure themselves, but he can't move.

Fuzzy and the Caseys are huddled together, stunned and in pain, but they're coherent. Suddenly, the curtains hanging across the galley entrance open and a passenger bursts in. He has a fully inflated life vest around his neck. 'I must get out of the plane. I can't breathe!' His eyes are wide and his face is flushed as he moves towards the rear emergency exit door. The life vest is riding high under his chin and neck; the inflated bladders restrict his breathing.

Fuzzy springs to action as best he can, yelling at the man to deflate his vest and get back to his seat immediately. But the passenger, staring blankly, shuffles towards the emergency exit as if in a trance. In his panicked state, he doesn't know how to deflate his vest.

It's up to Diana to take the initiative. Somehow, she reaches up with a pen in hand and inserts it into the inflation valve, manually deflating the vest. Then she slumps back to the floor. The passenger smiles sheepishly in relief. Fuzzy again yells at him to get back to his seat and fasten his seatbelt, 'Right *now*!' The passenger disappears through the galley curtains and returns to his seat.

The three injured colleagues exchange looks of disbelief at this almost-comical encounter. Then they think about the decision this passenger has made – he'd rather put on a life vest and jump out of the plane, despite its altitude, and take his chances on surviving the fall into the ocean, than remain inside the torture-cell that the cabin has become. If military-style ejection seats had been fitted to this Airbus for passenger use, they would all have been activated after this second dive.

Diana struggles to her feet, her shoulder still burning from the impact of the first jolt. Filled with a grim resolve, she positions herself between her husband and Fuzzy, and together they shuffle towards the crew seats. Fuzzy is secured first; she assists him to the right-hand crew seat, strapping him in with the seatbelt and shoulder harness. He's in agonising pain as he slumps forward against the straps.

Now Diana moves with her injured husband, shuffling towards the opposite crew seat. She gingerly lowers Peter onto the folding seat, then secures him with the full harness. He slips in and out of consciousness as he slumps against the restraining straps.

Diana remembers something: her airline emergency training stipulates that the escape doors must be manned in case they're required to be activated on landing – if we can land.

With raised voices, the cabin crew are now telling all the passengers to fasten seatbelts and move debris from the aisles, which must be clear in the event of an emergency evacuation. Oxygen masks dangle like jungle vines from the exposed stowage compartments in the aircraft's ceiling. Heroically, the crew don't let their own injuries and fears prevent them from performing their safety duties.

But they're aware of the terror of their situation – and it isn't over yet.

*

On the flight deck, the confused computers have completed their second psychotic takeover of the controls. I'm in the recovery phase again, probing the sidestick with small nose-up inputs until the aircraft responds. The plane is descending in a nose-low attitude, so I must again try to recover as smoothly as I can. We lose 400 feet of altitude in leveling off and commence climbing back to 37,000 feet.

By this time, my cage has been rattled. I've been sucker-punched twice, powerless to stop the aircraft's illogical and violent behaviour. How many more times will this happen? I can't and don't dwell on this question, but it's a critical situation when the captain of the plane becomes an ineffective observer.

I feel betrayed by the automation, which has twice taken control of the plane away from me. How can it fail in such a complex and aggressive way? I don't feel safe at all.

Something is moving on my chest above my left shirt pocket. I look down and am first shocked, then wryly amused to see my shirt fluttering. The fabric above my company wings is pulsating rapidly in synch with my pounding heart, and there's a drumming in my ears. I've never experienced this extreme heart rate before; I estimate it's at least 200 beats-per-minute. My life is at risk, so my body is responding.

'Hey, Ross, check this out.' I direct his gaze towards my chest.

His eyes flare in disbelief, and I realise that seeing your captain's heart beat this fast must be confronting. I flash an exaggerated expression at Ross, summarising the *what-the-fuck* feeling we are experiencing. It's a signal that I'm still functional enough to see the black humour in my body shifting into high gear.

To function effectively, though, I'll have to reduce that heart rate and calm my body down.

'Keep breathing, Ross!' I coax again, to both of us.

I'm doing the same; taking exaggerated breaths, holding them for a few seconds then exhaling. It's the drill of the professional athlete who's about to shoot a free-throw; I have to force myself back down from this adrenalin overdose. My skin is electrified; not ideal for this life-or-death battle with the computers.

As we're climbing back to 37,000 feet, the automatic pitch-trim wheel on the centre instrument panel isn't moving. Normally, this function is part of the computer's job of helping both the pilot and autopilot to fly with reduced force on the tail surfaces. The large black wheels next to each pilot's knee move in response to the computers feeding in corrections to the tail, and they should be moving now as I raise the aircraft's nose and climb again. But I notice a heavier feel in the control stick – turns out the computers have stopped helping me with this too. I check my primary flight display for the anticipated 'USE MANUAL PITCH TRIM' message emblazoned in red, but it isn't there. I manually move the wheel to relieve my stick pressure.

This is another cut in my knife fight. Each system fault and aircraft manoeuvre adds to my confusion. I'm being pushed further into a state of ineffectiveness. This pitch-trim issue is another distraction I don't need right now; I continue to swallow my frustrations and find a way to fly with the systems that are working at a basic level.

The stall and overspeed warnings keep blaring, and I suppress a sinking feeling that I'm trying to swim in a pool of quicksand.

I let Ross know we've lost the automatic pitch trim as more error messages are displayed. Again, the fault with the navigation module of ADIRU 1 is identified. We've already reconfigured that system but why is it active again? We skip it and move to the next alert message.

'Alternate law – protection lost,' Ross announces.

This message informs us we've lost the automated protection features of our flight-control system. These protections are computer-programmed boundaries that keep the aircraft in safe operation, despite the actions of the pilot, and assist if the pilot needs to perform extreme manoeuvres, for example, to avoid flying into a mountain. If these protections aren't available, then Airbus has determined that computer control is degraded from the normal operation. An error message and corresponding reconfiguration procedure are generated for the pilots to manage.

The procedural part of this fault is minimal. It's presented as a caution but there's nothing for us to do. I dismiss this error as not applicable to our situation, and Ross agrees.

I've already been thrust back into the traditional role of a pilot through the loss of most of the automation installed to make my job easier, and I will keep the aircraft flying safely as I have done throughout my career, with or without computer protections.

That's the gist of HAL's alert message … *I can't protect you anymore, Kev.* He's done a piss poor job so far.

I direct Ross to clear that message, then I quickly scan my primary flight display for visual confirmation of our Alternate Law state of computer control. There should be a series of amber-coloured 'noughts and crosses' displayed on the blue and brown background of the nose-attitude indication, but they're absent. A new question mark flashes before me: another false indication or another failure of the automation? I quickly store this snapshot into a subconscious compartment for further review when time allows.

My trust in the computer control is eroded to the point that I must question everything and this increases my workload further.

The warnings continue to bark at us through our speakers at maximum volume, reminding us that our knife fight is far from over.

Our next error message is more ominous.

Ross announces another failure in the operation of PRIM 3. There's an amber-coloured fault light on the PRIM 3 control switch.

*Why has it faulted again?*

'Reset PRIM 3?' Ross asks, his finger poised on the switch.

After our previous reset of PRIM 3, the aircraft pushed the nose down a second time. If we reset it again, will the computers be allowed to continue their violent manoeuvres?

Maybe we need to keep this primary computer in its faulted state until we have ensured this reconfiguration is safe or required for our landing.

'No – don't fuck with it,' is my instruction, and my decision.

Ross announces that the reconfiguration is complete and clears the centre screen.

The time is 1246. I've been on the flight deck for seven minutes. In that time, our world has been flipped upside down, and Ross and I are fighting for all our lives.

I need to reassure my passengers that their pilots are still alive, too. I hand over control to Ross and press the PA button. 'This is the captain. We are obviously having some problems with our flight-control system. Please remain seated with your seatbelts fastened. I'll get back to you shortly.'

Short and sweet, but what else can I tell them at this point? I'm being honest but I really don't know what is happening to us. I report what I see and let them know we are still working.

*

As my voice carries from the PA system throughout the cabin, the moaning, crying and sobbing immediately stop. It's deathly quiet as the whole plane listens to the short announcement.

Now they know the pilots are still functioning and still fighting! A small, collective sigh of relief drifts through the cabin. Their spirits are elevated, and their chances of survival are enhanced.

*

I notice my speed and altitude tapes are still jumping around and unreliable. My right thumb slides over to the manual disconnect button on the left thrust lever fitted to the central console; this removes automatic control of the engine's power. I push it, placing the engines into my manual control. It's also part of the unreliable speed procedure and now I have the time to action that step. The unreliable speed procedure is designed to remove the automation to prevent erratic operation. Could have fooled me.

'Manual thrust, Ross,' I announce to confirm my selection.

A new distraction adds to our confusion. Cleared of all the previous errors, the centre display is blank, but now the master caution chime becomes active:

*DING!*

The amber master caution light illuminates, directing our attention to the centre display screen. We catch a fleeting glimpse of an amber message as it flashes onto the screen – then, to our surprise, it disappears.

Immediately after, there's another chime, then another message that disappears, only to be replaced with another sequence of chime, flash of a message, disappearing act. These chimes add to the cacophony of the ever-present overspeed and stall alarms.

I didn't even get a chance to ask, 'Can it get any worse?', because it already has.

# 7.

Extreme flying and elevated body response are part of an aircraft carrier pilot's job description. Landing on the pitching deck of a ship, day or night, is a dangerous occupation that never wanes into the ordinary, even after a thousand traps. There are no margins for error; the precision required to survive carrier operations places every pilot on the edge of their mental and physical capacity.

Every naval aviator who's landed aboard an aircraft carrier has been exposed to the physiological phenomenon of fear-based body response, We conveniently categorised it as adrenalin release and we had to learn to cope with it effectively in order to survive. It has been recorded that a fighter pilot's heart rate is significantly higher while they're conducting a carrier landing than it is during the rigors of dogfighting. After a 'night trap', it would routinely take hours for our bodies to settle down enough from our high adrenalin levels to allow us to sleep. We referred to this post-landing condition as being 'wired'.

One of the most difficult landings I experienced occurred on my first deployment onboard the USS *America*, when I was flying the F-14. I was attached to the Fighting Aardvarks of Fighter Squadron 114 (VF-114) as part of Carrier Air Wing 11. My logbook notes the date as 1 July 1981, the month before we arrived in Perth. Our deployment had commenced in April, and my total carrier landings were approaching the 100 mark.

The carrier group operated in the southern Arabian Sea, an area we nicknamed 'Gonzo Station'. GONZO was an acronym for 'Gulf of Oman Northern Zone of Operations' and defined our operating area. This region was in the grip of the southern monsoon, with severely degrading visibility due to low clouds as well as gusty winds blowing mounds of dust from the Arabian peninsula out to sea. The dust mixed with sea spray, creating some challenging weather conditions.

On the day of my difficult landing, I was returning to the carrier from a routine combat air-patrol mission. In my back seat was 'Chain', my radar intercept officer: an experienced, second-cruise veteran.

We were instructed to enter a holding pattern and assigned a time to commence our instrument approach to the ship. The weather conditions had deteriorated significantly; our routine day landing now required an instrument procedure to deliver us to a point for a visual landing using 'the meatball'. Carrier (and training base) runways have an optical landing system that projects a precise visual glidepath to the pilot to guide him into the arresting wires of the ship. This is required because our landing area is constantly moving as the ship maintains its course and our landing area is so short. On earlier systems, the optimum glidepath was indicated by a glowing orange sphere, hence the nickname.

At the assigned approach time, I commenced my descent from 5000 feet. My target was to arrive at 1200 feet, 10 miles behind the carrier in the landing configuration of gear and flaps down. I'd then receive radar guidance from the carrier's air traffic control centre to a point less than a mile from landing. From there, I should see the ship and the angled landing area to complete the landing visually using the meatball.

This was the plan. But naval aviation is a constantly changing plan, so pilots learn to adjust and improvise to survive.

I entered a cloud layer that seemed to envelope the whole Earth as I descended below 1000 feet on my approach. I was flying on my instruments while following the guidance provided by the carrier-based precision controller. They are the best on the planet and would guide me precisely to my decision point.

It looked like my canopy was covered in frosted glass, a grey opacity with no sky or water visible. Only my instruments and the monotone instructions of the precision controller defined my position in space and distance.

Approaching two miles behind the ship, I started 'peeking' through the forward canopy to visually acquire the angled landing area. Normally, it was visible by this point, but I remained encapsulated in a grey blob of cloud. In my rear-view mirrors, I glimpsed Chain craning his head to the side in search of the carrier. We were getting closer to the ship but we couldn't see it. The precision controller continued to guide us.

One mile behind the ship, we were approaching the visual transition point. The aircraft was about 300 feet above the water;

I knew I should see the back of the carrier and the landing area centreline and the meatball. I didn't see anything.

Chain asked a quick question. 'See the ship, Sully?'

'Negative,' was my curt but precise response.

The controller made his last guidance instruction. 'Three quarters of a mile, call the ball.'

I was now required to transition to visually flying the meatball, and Chain would make a transmission confirming this, along with our aircraft side number and our fuel state. The landing signal officers (LSO) monitored this call; they were positioned on the left side of the landing area, providing guidance and safety to landing aircraft.

We were on hot mic and Chain asked, 'Sully?' as we progressed past the ball call position.

I hesitated for a tenth of a second, hoping the ship would miraculously appear, but I didn't see anything. It was up to me to make the next call. 'Clara.' That meant: 'I need "clarification".'

I didn't see the ball. I didn't see the ship.

I didn't see anything!

It would be the only time in my carrier career that I uttered this confession of blindness to the LSOs. They would have to guide me once I made this call.

Simultaneously, I realised that, based on my rate of descent and tracking, I'd have to abort my approach in the next second – I was drifting to the right of the ship, where the rescue helicopter orbited, and getting dangerously close to the ocean.

As my hands tightened on my engine throttles to initiate the 'wave off', the LSOs transmitted, 'Bring it back hard left, Sully!'

They could see me from their position, and, instinctively, I slammed my control stick to the left. Immediately, the deck and the meatball appeared, so I transmitted, 'Ball.' Then, full stick to the right to align with the centreline, keeping the meatball centred, I landed with full force of a carrier landing. My hook caught an arresting wire as I added full power and welcomed the violent deceleration as the wire slowed me down.

I had to quickly clear the landing area, but my legs were shaking considerably as I displaced the rudder pedals and tried to follow the directions of the flight-deck marshaller. We were parked right next to the angled landing area and had front-row seats as the remaining aircraft attempted to land.

Chain piped up from the back, 'Good job, Sully! Check out the mast!'

The ship's mast was 60 feet above the level of the flight deck, and the flight deck was 60 feet above the ocean. The top of the mast was obscured with cloud, which meant the cloud layer was about 100 feet above the ocean – 100 feet below our landing limits! I'd just completed a manual approach well below our landing minimums, which probably explained my bad case of 'machine-gun legs'. I hadn't experienced this type of extreme body response before.

We retired to the squadron's ready room, my sweaty face still bearing the red outline of my oxygen mask around my nose and mouth. Soon the LSOs entered solemnly, like a rugby scrum of white-vested clerics searching me out. Their hair was dishevelled and their faces greasy from the wind, sea spray and desert sand that had been pelting them for the past forty-five minutes on their exposed platform. They scrutinised me as the Air Wing LSO delivered his debrief. My extreme manoeuvring earned a 'nice correction' comment and a grade of 'okay', meaning 'good landing', a small reward for getting aboard from an extreme position.

As I listened to their assessment, the closed-circuit TV in our squadron ready room was replaying the recovery. My extreme manoeuvres were there for all the ship's company to see. I took a long drag on an imaginary cigarette. 'Thanks for getting me aboard, Paddles,' I offered, after an exaggerated exhale. (LSOs earned their nickname in pre-meatball days when they used two brightly coloured paddles to give visual instruction to landing pilots.) With a few wry chuckles and grunts, the scrum left in search of other pilots to debrief. I slumped back into my chair and mentally flew the approach over and over again as I watched the CCTV; my recovery was replayed several times. I was still sweating, the ship's air-conditioner was too feeble to help me cool down and my flight suit material was too dense to allow any fresh air to touch my clammy skin. My heart was still racing, but I was enjoying the buzz from the chemicals my body was providing. I felt elated and energised.

I mentally patted myself on the back for getting aboard on such a difficult day, then stored this new physical sensation in my Filofax of memories. Crazily enough, I never really felt threatened with death in my military career.

We commonly refer to these threat chemicals as adrenalin, but there's more than one ingredient; it's a complex mix of hormones released by the body's glands to prepare you for a threatening event.

QPA was providing plenty of stimuli.

*

'Emergency cancel, Ross.'

Below the centre screen in the cockpit is a small panel containing an emergency cancel button. It's there to cancel nuisance alerts and noises associated with non-routine events, which the pilot determines to be false. Well, all of them are false, and now we're getting even more of them through the dinging of momentary error messages appearing and disappearing on our screen.

Ross pushes the emergency cancel button. He pushes it again. And again.

The dings and stall alerts continue. We aren't allowed to silence them because this system, too, is computer-controlled and confused.

Great.

I feel cheated, like I've just driven out of the caryard with a newly purchased, state-of-the-art automobile that's, in fact, a lemon. I believed the salesman's glowing pitch and handed over my hard-earned cash, and the fine print in the warranty says they won't fix it. I'm far from impressed; in fact, I'm furious.

Silently, I review the condition of my high-tech A330: manual rather than automatic thrust; manual pitch trim; manual flight; aircraft behaviour that's outside the realm of our training and knowledge base; and a cockpit flooded with confusing warning and caution alarms that can't be silenced. I'm flying through a high-tech computer, but which one is moving the controls? It could be any one of the three primary computers or two secondary computers, but this information isn't displayed to me.

I have other serious issues to factor in: a planeload of passengers who have probably suffered injuries, especially after the first pitch-down, and we're still two hours' flight time to our destination, Perth. I'm starting the think we may not get there.

Add to that, the two on-duty pilots are trying to operate through the buzz of adrenalin overload after facing the probability of flying into the ocean under rogue computer control. We are wired, but this isn't the good buzz of a daring carrier landing – it's inhibiting our ability to analyse the situation and perform minute tasks.

Our altitude has stabilised at 37,000 feet again. My sidestick inputs generate a twitchy response from the aircraft now. I must make smaller, lighter inputs to keep things smooth. Is this a consequence of Alternate Law? Another distraction thrown into the complex equation I'm trying to solve.

We have to get our first officer back in the cockpit to help out, so I ask Ross to make contact again through the interphone.

*

In the forward galley, Pete and Lisa have survived the second pitch-down. The cabin interphone call from minutes ago is still active, the handset dangling by its cord. Lisa tries to make an all-stations call to her crew to find out their status; no one answers. She resets the handset, frustrated that she can't reach them.

The interphone rings a second later; it's Ross calling from the flight deck. 'Tell Pete he can come up now! We'll watch him and open the door.'

Lisa acknowledges, but doesn't really want the first officer to leave her. 'Pete, you can go up now. They're waiting for you!'

Before he makes his dash, he straps her into her crew seat next to the galley. Then, like Usain Bolt out of the blocks, Pete sprints towards the cockpit door, picking his way around fallen luggage. He hears the electrical bolt of the cockpit door unlatch and pushes it open.

He's back on the flight deck, slamming the door behind him – but he isn't prepared for the paper charts and folders littering the floor. He almost loses his footing as he scrambles into his seat, which Ross vacated seconds before.

*

'It's a shitfight out there,' Pete says as he straps himself in. 'I think I broke my nose.'

I look over at him; his eyes are wide open, trying to take in as much information as possible while he adjusts to being back on the flight deck. The bridge of his nose is reddening.

'Congratulations,' I reply in an exaggerated tone, like he just won a crappy prize on a gameshow. 'Strap in, we're in trouble.'

With those words – words I've never even contemplated uttering in any previous emergency situation throughout my entire extreme flying career – I've aptly summarised our position.

The stall and overspeed warnings welcome Pete to our own shitfight. I quickly let him know that the dives were computer-generated and I couldn't stop them with my control-stick inputs. I then state our roles for the remainder of the flight: I'll remain as the active pilot, and Pete will assume the support role.

'Declare a "mayday"? Learmonth?' he prompts, adding that he observed passengers and crew with injuries. Pete is on the ball.

I've already contemplated the divert decision. Our aircraft is in a state of systems meltdown, and I have no real choice but to land as soon as possible. Thanks to air-force requirements, the Learmonth runway is long enough for an A330. The civil facilities are modest – typical of those for a small regional airport – but our company has certified that Learmonth is suitable for use as a diversion airport, and it's definitely right for us now. Continuing on to Perth could expose us to further pitch-downs and more passenger injuries.

I modify Pete's first idea. 'No, Pete, let's start with a "pan", not a "mayday". Degraded flight controls and injuries in the cabin.'

A pan declaration is the international urgency signal to communicate a degraded flight situation that doesn't pose an immediate danger to anyone's life or the aircraft at the time of transmission. A pan call is an alert to air traffic control and rescue agencies that a serious issue exits, asking them to prepare to provide assistance. It's pilot-talk for 'Hey, we've got a problem and we'll get back to you.'

Mayday is different; there's a serious threat to life and to the continued safe conduct of the flight. Emergency assistance is required. I need more information before I can justify an upgrade to a mayday.

Ten minutes since this began, it's time to let the world know we need help.

Soberly, I've already concluded that we're heading into mayday territory now.

# 8.

Pete makes his pan call to Melbourne Air Traffic Control: 'Melbourne Centre. Pan-pan. Pan-pan. Pan-pan. Qantas 72. Qantas 72. Qantas 72. Experiencing flight-control problems and we have injuries in the cabin. Require diversion to Learmonth and tracking direct to Learmonth.'

'Qantas 72, Melbourne Centre. Pan acknowledged. Track direct Learmonth. Cleared to descend to 35,000 feet.'

I know that the controller will be scratching his head, trying to make sense of the blaring background alerts that accompany our radio call for help.

'Ross, try to get a damage report from the cabin.' We need more information about damage and injuries, because elevating our emergency condition to mayday requires more justification.

I use this short break to give Pete a summary of our current situation. He nods as I describe the degraded, manual manipulation that I've been forced to accept.

I direct him to start our diversion phase to Learmonth. For us, it's a rarely used contingency airfield; none of us have been there before.

Well, that's not exactly true. A short visit to this airport and to the Harold E. Holt naval base was the start of my path to live in Australia.

*

After my first visit to Western Australia in August 1981, an opportunity to return came in December that year when we'd completed our deployment. Two of my squadron mates were travelling back to Perth for the southern summer and asked if I wanted to go, too. We'd hitch a ride on military transports if any seats were available.

It just so happened that a transport service was scheduled between California and Learmonth, and we were on it.

After a few days of rough travel in the back of a Lockheed C-141 Starlifter, we landed in Learmonth. We stayed the night at Harold E. Holt naval base and flew to Perth the following day on Airlines of West Australia. The blowflies were particularly persistent and fearless at Exmouth. We made the mistake of playing a set of tennis but most of our ground strokes were reserved for the blowies.

Our holiday in Perth was a relaxing tonic after eight months at sea. I felt comfortable with the Aussie beach-based lifestyle, and the laidback approach to living was a stark contrast to life at a military base in Southern California. I hoped an opportunity would present itself for me to spend more time here.

That opportunity came during my second cruise on the 'Big E', the USS *Enterprise*. The Royal Australia Air Force advertised a 'personal exchange' position: the Australian government was buying the F-18 Hornet to replace the ageing French-made Mirage 3 aircraft as their frontline fighter and it is common practice for militaries in the Western world to 'exchange' personnel and operational doctrine. My Aussie counterpart would be based at NAS Lemoore in California and would learn to fly the F-18 as an instructor pilot.

I was due to transition to a shore-based position; luckily, this exchange assignment would satisfy that requirement. And my experience fit the stringent RAAF job prerequisites: junior officer, operational experience, instructor experience and fighter weapons school graduate. This was a full house of military experience that few other applicants would have, so I applied.

I didn't hear much from the Pentagon after that, but my life changed in another way – I became engaged to a young woman from Perth whom I'd met the previous year. The exchange assignment would help those marriage plans become a reality, and my feeling was that instructing in the RAAF would be a slow-paced assignment compared to the hectic life of a navy pilot. But though my aircraft-carrier cruise was drawing to an end, I still had no word on my next assignment.

On 27 April 1983, Carrier Air Wing 11 completed their second deployment, and it was time for all the squadrons to fly off from the USS *Enterprise* and return to their home base. Fighter Squadron 114 was based in San Diego at NAS Miramar, and our group of F-14s would arrive in a mass formation of sixteen aircraft.

I almost didn't make the fly-off.

I was assigned the aircraft that our maintenance division had designated the 'hangar queen'. Each squadron on the ship kept one aircraft in the hangar to use as a source of spare parts for repairing their other jets; towards the end of the cruise, the hangar queen was rebuilt in preparation for the flight home. I was a qualified post-maintenance test pilot in my squadron, so the job of flying this jet off the carrier was left to me, even though it wasn't in the greatest shape. It was still being worked on and inspected as I waited to fly out. Eventually, mine was the last aircraft to be shot off the catapult, and I said goodbye to the 'Big E'.

The VF-114 Aardvarks had been away for eight and a half months, and the Miramar flight line was crowded with friends and families to welcome us home. Not only was our fly past spectacular, but, once again, I was positioned in the middle of the formation. Straight after landing, the squadron stopped and lined-up on the taxiway. We pulled off all our flight gear in the small confines of our cockpits down to our orange flight suits and soft khaki hats, ready to join the party.

But after landing, I started having problems with my Tomcat; it still thought it was in the air and wouldn't let me steer with the nose wheel. I limped into my parking spot as much as my failing jet would allow.

As I shut down the engines and opened the canopy, a senior officer climbed up to the cockpit, gave me a rose and told me I was going to Australia!

I was awarded the assignment as the first US Navy exchange pilot to the RAAF. One month later, in May 1983, I arrived at the base in Williamtown, New South Wales, as a fighter combat instructor. The future leaders of the RAAF would be my students, and I would be married later that year.

This posting proved to be the start of a new life and a new adventure in Australia. In 1986 I resigned from the US Navy and was hired by Qantas.

*

It looks like today I'll set foot on the Learmonth tarmac once again, but in radically different circumstances. We agree that diverting there is the safest and smartest course of action. Under my direction, Pete dials in a slower speed, followed by our cleared altitude of 35,000 feet.

We are on our way.

I'm already pointing the aircraft to Learmonth visually; we're 80 kilometres to the south of the airfield, and we can see its shape. Cut out of the burnt-red earth of Australia's wild outback is a long grey strip of concrete.

Pete's competence as an experienced ex-military pilot is comforting. I know he'll be able to keep up and provide solid support with this high-tempo emergency workload. Ross, likewise, is working professionally through the haze of the pitch-down's aftermath. We know our position is still precarious and the reality is that there is more work to be done to land safely. We work with quiet professionalism, and it's gratifying to know that I don't have to monitor their performance, in stark contrast to the supervision directed at our deranged Airbus.

Learmonth is an unknown airport for us so we'll need some time to prepare. I've selected minimum speed to give us some time, and I must prepare a strategy to address landing an aircraft that's behaving very badly. This scenario isn't covered in any of our manuals.

I muse privately that the shit must be hitting the fan at Airbus headquarters by now.

Pete is struggling to insert a 'direct to Learmonth' command into his computer, which would allow us to access the appropriate digital charts for our approach and landing. He tries several times, but this computer is also refusing to interact. Another element of our automation has failed, forcing us to revert to basic visual navigation. Fortunately, we can clearly see the airport, so I fly towards it as Pete backs me up with his target selections.

'Geeze … okay, Pete,' is all I can muster as I watch Pete's finger continuously jab the 'insert' button in frustration.

I continue to crosscheck the standby instruments and Pete's instrumentation to confirm our safe flight path. I work out that the yellow guidance bars superimposed on my primary display are working and in agreement with Pete's. They'll give me another visual cue to assist in following the performance targets he's setting.

We need to start our descent now and move into the next phase of our procedural sequence for landing.

Pete signals our descent, 'Qantas 72, left 37,000.'

'Qantas 72, roger. Clear to leave controlled airspace.'

Melbourne Centre has cleared us to leave their area of controlled airspace that projects upwards from 25,000 feet over the Earth. We aren't in radar control; this will add to our workload.

*

Lisa is still strapped into her crew seat. Since the second pitch-down, the aircraft has stabilised, but the many injured passengers are in various states of pain and discomfort. She glances over at her two crew, Samantha and Jen, now seated opposite her on the right side of the galley. Their concerned eyes briefly lock before they're interrupted by shouts in the cabin.

They're shocked to see that Rory, another flight attendant, is out of his seat and running through his area, bravely assisting the injured as best he can while yelling instructions to others to fasten their seatbelts. He enters the forward galley, limping, then turns back into his area of responsibility and completes his loop. Lisa, Samantha and Jen scream at him to return to the safety of his crew seat and buckle in. He says he must check on the babies secured in their bassinets on the cabin wall behind the galley. In a flash, Rory disappears back into the cabin, offering support and instructions to his passengers, then he hobbles to secure himself in his seat.

The interphone rings – it's Ross making contact with Lisa to ask for a damage report. From her seat she can't clearly see down to the back of the aircraft, so she reports that her vision is blocked.

Another voice now speaks on the interphone. 'There are severe injuries in the rear of the aircraft! Broken bones, lacerations and maybe spinal injuries! There's damage to the cabin and the ceilings!' It's Diana, reporting from the rear galley.

Through the buzz of our hectic workload, Pete and I hear this report through our headsets. Our eyes lock, and simultaneously we come to the same decision: our emergency status is justified.

'That's it,' I say.

'Mayday?' Pete prompts.

I confirm our situation. 'Mayday.'

I can't risk flying longer than necessary with a cabin full of severely injured passengers and crew, and I'm sending out the call that we need all the help we can get. Added to that is the uncertainty generated by the erratic behaviour of the aircraft and our computers systems.

I need to get on the ground as quickly as possible, but I haven't figured that part out yet.

*

Diana has used her initiative by responding to the all-stations interphone call. Her damage report provides critical information to the pilots, and now she returns to the seat assigned on her boarding pass, her progress impaired by debris.

An elderly man makes eye contact with Diana as she hurries past. His face is streaked in blood and his scalp is badly torn. Part of his skull is clearly visible. Grabbing a small pillow, she presses it firmly on his bleeding head and gestures to him to keep it in place. He nods in understanding, and his eyes glisten in gratitude.

When Diana arrives at her seat and belts in, she notices that seated next to her is a young woman gasping for breath and in distress from a panic attack. Diana tries to retrieve one of the dangling oxygen masks to help the woman breathe, but it's too high up. She holds the woman's hand, hoping to comfort her.

In the row behind Diana's right shoulder, two young sisters under the age of ten are travelling alone. They are distraught but their despair is barely audible among the moans cries of the many panic-stricken, injured passengers in the rear of the plane. She reaches out to them but they're inconsolable. To distract them, she sings 'Row, Row, Row Your Boat'.

Secure in her seat, again she directs her gaze towards some activity in the front of her cabin zone. Horrified, she sees a pair of limp legs dangling from the ceiling. She can't see the head but she can see a hole from which the unconscious woman's body hangs like a gallows victim. Passengers' arms are outstretched around her, trying to free her.

*What a nightmare*, Diana is thinking.

Then her thoughts focus on her two daughters, seated in the front of the plane. *Are they okay?* She prays they kept their seatbelts fastened, but her heart and mind are heavy with worry.

Although her injured shoulder is aching terribly, she continues to hold the hand of the panicked woman next to her. The young girls are singing now through their subdued sobbing.

Peter and Fuzzy drift in and out of consciousness, secured in the crew rest seats, while Diana silently prays the plane won't repeat its violent behaviour and that this horror flight will end soon.

# 9.

Pete has been back on the flight deck for three minutes. In that time we've declared a pan, started our diversion and decided to declare a mayday.

Before Pete presses his radio switch, he recalls a saying from his old Australian Navy squadron: 'If you're gonna make a mayday call, don't sound like a dick!'

'Melbourne Centre, mayday. Mayday. Mayday. Qantas 72. Qantas 72. Qantas 72. Flight-control problems, and we have significant numbers of serious injuries onboard. Tracking direct Learmonth.'

'Qantas 72, Melbourne Centre. Mayday acknowledged. Track direct Learmonth.'

With the mayday declared, the shit has really hit the fan.

Melbourne Air Traffic Control pushes the 'crash button', figuratively, and a nationwide alert phase is activated. All airborne commercial aircraft listening to Melbourne control frequencies in the southern half of Australia hear this distress call because it's automatically retransmitted through a repeater system.

Pilots in all those flight decks will be looking at each other incredulously. *What the bloody hell has happened to Qantas 72?!*

A colleague and good friend is one of those pilots, flying between Perth and Sydney. 'That could be Kev Sullivan,' he says to his first officer, knowing that I'm flying today from Singapore. They listen intently to their radios as they fly east.

I retard the power now to descend gently at a reduced speed. We still need time to prepare for our approach and to familiarise ourselves with the Learmonth airport charts, and we'll need paper copies. The warning alerts and caution chimes continue.

As the thrust is reduced, there's a noticeable change in cabin pressurisation.

Ross's ears are popping. He quickly confirms that the automatic control has failed and the pressure is fluctuating. It may already have identified itself as a faulty system, but the constant scrolling of caution messages would have hidden this malfunction. It's another critical system that's offline adding to our heavy workload – we must manually control the cabin pressure as we descend. Ross volunteers to do this.

When Pete backs up Ross's troubleshooting of the cabin pressure system, he selects the appropriate systems page and sees that it confirms failures of the pressure system's computers: they're highlighted in amber to indicate the fault. As a matter of course, Pete selects a summary page marked 'status'. He's blown away by the huge list of inoperative and faulted systems in the fifteen minutes since this began.

'Hey, Kev,' Pete says, his finger prompting me to look down at the screen.

There are two pages full of faulted systems, including some significant failures that will affect our landing. It's like a very disappointing report card for a parent to read, full of failing grades and unacceptable behaviour.

I don't have time to handle these failures now. I acknowledge the status information, and Pete tells me about the loss of the automatic braking system. Mentally, I reshuffle my priorities to move that consideration to a higher level.

I'm pointing our disabled jet towards the runway, and I have to keep things simple. We are all struggling to concentrate through the haze of fight-or-flight chemicals. I will fly over the airfield once we get there and circle as we continue our deliberate descent.

What will fail next on our A330? These system faults have not been displayed to us, lost in the merry-go-round of cycling master caution alerts. I don't even have time to smirk or bitch now. We're down to the level of a small propeller-driven aircraft. The workload and distraction level required to manage our situation is escalating as we get closer to Learmonth.

I'm relying on 'Aviate, Navigate, Communicate'. This helps me prioritise my many tasks and decisions while I manage our emergency situation. In the extreme and distracting environment our confused computers have created, this mantra will be my lifeline if I can maintain my discipline and concentration.

I grab this small opportunity to discuss the sequence of failures with Pete. He listens carefully, his finger selecting the various systems pages on the centre screen as he searches for something we may have missed.

He suggests a call to our company maintenance unit, a 24-hour 'assistance hotline' based in Sydney. The aircraft is constantly streaming systems information and flight parameters through a data link to our home base. What if they can see something we can't access?

I agree, and Pete calls them on our satellite phone. He continues to back up my manual flying with target selections while he communicates with the maintenance unit.

We are calm and methodical. It's up to me to manage the pace and workload as we descend towards the runway. I have only two hands and don't hesitate to delegate responsibilities to my two trustworthy pilots. Because we can't access the digital airport information in our navigation computer, I assign Ross the task of finding the paper charts for Learmonth and to prepare an airport briefing for us.

This is no easy job because the paper contents of this manual have exploded on impact during the first pitch-down. Ross must sift through the scattered papers to find the right charts. He's also manually controlling the cabin pressure and maintaining communication with Lisa in the cabin.

Using his initiative, he activates the airfield pilot-assisted lighting on the runway through one of the radios. This will give us a lighting aid that we'll use for landing. The lighting system at Learmonth is normally off unless activated by a pilot using the airfield; there's no control tower or radar control at this small airport.

The flight deck workload for all three pilots is ramping up. I think we are busier than all the traders on the floor of the New

York Stock Exchange at the opening bell. We steal short scans of each other as we concentrate on our individual tasks.

Ross, Pete and I have 'split up' temporarily as we work independently, but we're all listening to the radios, the cabin interphone and Pete's satphone. We're all silently reciting our mantra of Aviate, Navigate, Communicate. Our workload is enormous, yet Ross and Pete are each monitoring my control of the flight path. They're on alert for opportunities to absorb tasks that will alleviate some of my workload, ensuring my manual flying is safe and appropriate.

I assume responsibility for communications on the primary radio frequency, Melbourne Centre. They're in charge of this airspace and for assisting us in this emergency phase and they're calling me now.

'Qantas 72, Melbourne Centre.'

'Go ahead.'

'Qantas 72, emergency services need 30 minutes to prepare for your landing at Learmonth. Adjust arrival time accordingly.'

'Roger. We will need at least 30 minutes to prepare, too.'

'Copy Qantas 72. Say souls on board and any dangerous goods.'

'Three hundred and fifteen souls on board. No dangerous goods. Qantas 72.'

The air traffic controller will share that information with all the agencies being mobilised to help.

There's a brief moment now for me to take stock of our situation. I must formulate a strategy and an arrival plan that's simple and effective, but that also addresses our flight-control issues and cabin injuries. The systems that are working are enough to form the first part of my arrival plan but I have a ways to go to fill in my strategy.

As my brain runs through the loops of possible solutions, one question remains: will the aircraft command a pitch-down at low altitude?

The aircraft is hiding information from me that I desperately need. I still don't know the reasons for the two previous computer-generated manoeuvres or why the computers continue to generate the erroneous stall and overspeed warnings along with the continuous caution chimes. I must assume the worst and consider the possibility of more bad aircraft behaviour at lower altitude.

Our situation is critical and nothing in my commercial experience and training addresses our complex predicament. I fall back on my military training to help me plan for this most extreme and unforeseen emergency.

# 10.

Landing a plane on the deck of an aircraft carrier is difficult, delicate and terrifying to some, but for many it's the most fun you can have with your clothes on. It tests a pilot's ability to fly their machine precisely and intricately while controlling their body and mind to land on the deck of an aircraft carrier. It's the most dangerous evolution you can perform in an aircraft.

Early in their training syllabus, a navy pilot learns to 'fly the meatball' and use angle of attack instead of airspeed. Optimum angle of attack is maintained through the use of a small gauge on the instrument panel that looks a little like traffic lights; the green, yellow and red symbols are used to indicate speed and therefore body angle to the pilot. The correct body angle is a critical alignment of the arresting hook relative to the aircraft's flight path to ensure an arresting wire is snagged on landing.

There are no features on the ocean's surface to indicate if you're positioned correctly as you approach the aircraft carrier, so the training concentrates on developing a rapid instrument scan. There are specific altitude gates to fly through to deliver you into the 'groove' for the final approach and landing, and these are accomplished mostly on instruments until you transition to the visual portion to fly the meatball to landing. Each approach is a precision manoeuvre with the objective of accomplishing a safe landing on the moving platform of an aircraft carrier – and it's always moving.

Day carrier qualifications were completed on both the T-2 and TA-4 during basic and advanced jet training. These qualifications were completed solo, meaning without an instructor pilot in the back seat. The landing signal officers (LSOs) told us solemnly that no one was crazy enough to go to the ship in a student's back seat on their first time out. Our LSOs warned us, as they'd been warned before their first carrier qual, that everyone on the ship was trying to kill us. To survive this day, we must trust no one, always be a step ahead and anticipate any instructions. This would be no picnic.

They showed us a movie, *Carrier Mishaps*: a preview of the serious nature of our task. Broken aircraft covered in flames crashed onto the deck or into the sea, to be run over by the carrier. A watery violent death was the fate for those unprepared.

My logbook shows I completed sixty-two simulated carrier landings at our base in preparation for the qualification. Every landing was graded, my performance under the intense scrutiny of the LSOs. Finally, I was cleared for the real thing.

The date of my carrier qualification was 12 May 1978. I was twenty-two years old.

Underway in the Gulf of Mexico was Training Aircraft Carrier 16, CVT-16, the USS *Lexington*, waiting to welcome us aboard. The 'Lady Lex' was a combat veteran from the Pacific campaign of World War II and still sported her original wooden flight deck. In the late 1950s, she'd been remodelled with an angled deck and designated as a training ship.

This was to be a special day, a baptism into the world of carrier aviation – a world of flame and heat and speed and danger – a world where I yearned to belong by the end of the afternoon.

None of us neophyte naval aviators slept well the night before. We were going out to the boat for real. The meatball would be attached to a real ship, with real arrestor cables and real steam-driven catapults.

A group of the instructor pilots were tasked to lead flights of three students ('chicks') out to the 'boat'. The qualification sequence was briefed again: two touch-and-go landings with the tailhook up and, when ordered, extending the hook to complete four arrested landings (traps) and four catapult launches (cats). The lead instructors would drop their chicks into the landing pattern, complete a touch-and-go, climb overhead the ship, wait for their chicks to complete or fail their qualification, then lead us home. Our base was Chase Field, Naval Air Station Beeville, Texas.

We suited up and prepared for take-off. It wasn't quite summer, but all the students were sweating profusely. I'd adorned my helmet with two green shamrocks cut out of reflective tape, and I would need their luck today.

Our flight of four Buckeyes lined up in echelon on the runway; I was Number 4, and my jet's side number was 605. At five-second intervals, we took off individually and quickly rendezvoused with our flight lead, 'Hondo'. A US Marine Corps pilot with combat experience in Vietnam, Hondo was usually a hard-ass instructor. But because we were senior student pilots about to be initiated into carrier aviation, he cut us some slack. He nodded in approval at our join-up, then jabbed his thumb over his shoulders, left and right, instructing us to move into a loose cruise formation as we climbed and moved out towards the Gulf.

When we crossed the hazy southeast coastline of Texas and went 'feet wet', I mentally prepared for what lay ahead. With a few deep breaths I confronted the fact I was going to do this.

Our flight of four checked in with 'Strike', the *Lexington*'s air traffic control centre. We were instructed to enter a counterclockwise holding pattern above the ship, and we maintained our loose formation as we circled. I stole glimpses of the orange-and-white T-2s landing, taxiing and being catapulted off the deck. It looked like a mechanised toy set, but soon it would be my turn to descend into it. I removed my oxygen mask briefly and wiped the ever-present perspiration off my face and around my mouth with the back of my Nomex glove.

After thirty minutes of circling, our time had come; we started to close up our formation in anticipation of entering the pattern. Everything we did now as student aviators was on show as we built reputations that would follow us through our careers.

We were descending. I sensed the water getting closer to my jet, but I was too busy maintaining my position in formation to look around. Hondo checked in with the carrier control area. A gruff voice acknowledged and answered abruptly, 'Charlie.' This was the code word for us to enter the traffic pattern. The deck was ready for us.

The gruff voice belonged to the 'air boss' of the *Lexington.* This was the air officer: a veteran, senior naval aviator manning and controlling the traffic pattern from atop the carrier. It was a stressful billet for anyone to complete, and here came three more 'studs' to make his day even more stressful. He's portrayed in the movie *Top Gun*, spilling his coffee as Maverick makes his high-speed passes and buzzes the tower.

The boss transmitted the ship's course, which we'd use to maintain position in the landing pattern, then growled at one of the students on the flight deck for using too much power while taxiing. On the deck, an aircraft using too much power could easily blow a sailor overboard with their jet blast.

Hondo clenched his fist and pumped it up and down, the signal for us to assume 'echelon right' formation for our entry into the pattern. From where I sat in Position 4, the T-2s were rocking up and down, aligned with our flight lead as we descended and rolled out at 800 feet behind the ship. I breathed slowly and deeply, trying to control my excitement, fly smoothly and maintain position as we approached from astern.

The boss cleared our flight to enter the pattern. We were flying past the *Lexington* now, and I glimpsed the white '16' painted on the bow of the deck. Our Buckeyes were bouncing in formation as we waited for Hondo's signal; he looked over at Number 2 and blew him a kiss with his right hand before racking his jet into an 80-degree left bank. This wasn't a term of endearment but the signal that the lead was breaking from the formation. Number 2 was now the lead; he'd started his clock, waiting for his forty-five seconds to elapse before he kissed us off and the belly of his T-2 disappeared behind us.

After Number 3 broke, I came onto my instruments, maintaining wings level, and flew the ship's heading on my compass. I started my clock and waited for my forty-five seconds to run before I broke into the landing pattern: the longest forty-five seconds in my young life.

As precisely as the Buckeye's roll rate allowed, I smartly rolled into an 80-degree angle of bank, pulling my throttles to idle, and extended my speed brakes. Holding 800 feet, I used the g-force and speed brakes to slow me down to the gear and flap extension speeds. Gently, I descended to 600 feet, adjusting my track by flying the ship's reciprocal heading to achieve the proper 'abeam' point and then extend my landing gear and flaps. Oh, and lock my harness! My first landing would be a touch-and-go, so my hook stayed up.

'605. Sullivan. Four point five.' My fuel state of 4500 pounds and my name were equally critical in this initial report. Each student's fuel quantity was closely monitored, and we had to be identified for the qualification.

I was manually trimming the wings and nose of the aircraft on downwind to balance the aircraft flight path, and I removed my hand briefly from the control stick to check that my jet was in a hands-off state of trim.

As I started my approach, I was abeam the LSO platform, positioned on the port side towards the stern. Using the clock, I manipulated my rate of descent, speed and angle-of-bank markers to help me fly to the 90-degree point. I was locked into the Zen of the carrier approach. My scan worked at maximum speed as I checked my progress, then I quickly glanced up and looked at the ship's wake, which I'd pass over to align myself with the 10-degree offset landing area.

The LSOs on the deck barked at me to keep turning my aircraft; unconsciously, I'd reduced my angle of bank as I looked up at the centreline, a common mistake.

Passing over the white and turbulent wake of the Lex, I lined up with the centreline and saw the meatball. Smoothly adjusting my power and nose attitude would keep the ball centred and deliver me into the wires for a trap. Also, this was critical in maintaining the optimum angle of attack to ensure correct body angle for my hook to catch a wire – too fast and my hook might skip over it, and I would 'bolter'; too slow and I'd catch a wire closer to the stern, and the LSOs would get quite vocal. My amber 'doughnut' in the middle of my AoA gauge showed that I was 'on speed'.

Every pitch, roll and power input required a counter-correction, and so it went to landing – a constant series of counter-corrections until the wheels hit the deck. For the optimum approach, they needed to be small, smooth inputs to the stick and power, but adrenalin could easily overcome this desired performance.

I heard the LSO issuing instructions to the jet in front of me:

'Get it back on speed.'

'Don't settle.'

'Power. *Pow-weeerrr!*'

As I rolled the wings level into the groove, I was rewarded with the vision of a centred ball and the centreline of the angled deck nicely aligned with my jet. I tried to maintain focus as I glimpsed the T-2 ahead, climbing away after his touch-and-go. Time to make the first ball call of my career.

I confirmed that I could see the meatball and that I was flying a T-2 Buckeye, and reported my fuel state again: '605, Buckeye ball, 4.4.'

'Roger ball,' came the LSO's laconic drawl.

I wasn't looking at the deck or the ship or the surface of the sea. I was scanning the centreline of the angled deck. The chant of the naval aviator had started: 'Meatball, line up, angle of attack,' which set my priorities. I flicked the wings to the right as I tracked my changing centreline. The angled deck was constantly moving to the right as I descended. I was 'in the middle now' on my approach, altitude less that 200 feet. The chant changed to the priorities of 'Meatball, line up, meatball, line up'. By now, speed and angle of attack needed to be stabilised but still required correction in concert with the approach angle and centreline tracking.

The LSOs are visually scanning our aircraft geometry to ensure our approach angle is precise. An incorrect speed changes the angle of the hook relative to the deck and the landing area. Almost every student was instructed at some point to 'get it back on speed' by the LSOs. All aspects of a carrier approach are critical.

'Little power,' coaxed the LSO, and I confirmed my angle-of-attack indicator was showing a slightly slow condition. I gently 'walked' the throttles up a fraction to satisfy the low-energy situation, then recorrected.

I progressed to the 'in close' position as the deck filled my canopy in my peripheral vision. I was still chanting as the ramp – the tail-end of the deck – slid past my jet. My chant was rapid now, my priority was 'Meatball-meatball-meatball', my vision locked on the meatball as my wheels hit the deck. The ball was slightly high as I selected full power and shifted my eyes forward to complete my touch-and-go.

The island of the carrier was a blur as I was off the deck and flying again.

'605 turn downwind now,' barked the air boss. There must have been a gap in the landing pattern spacing and he was using me to close up the landing pattern.

'605, rog-*gaaa*,' I replied, cranking my Buckeye left to join the pattern.

With a touch-and-go under my belt, and still alive, I felt mildly confident. Mentally, I reviewed my last landing as I turned. Maybe I'd used a bit too much power while correcting to the LSOs' instruction; next time I'd ensure my power corrections were more precise and my anticipation more focused.

*Fly the ball!* I reminded myself.

I completed another touch-and-go without much comment from the LSOs and turned downwind. I knew what was coming next.

'605, hook down,' came the boss's instruction. The time had come for my first trap: my first arrested landing on an aircraft carrier.

'605,' I responded, as I slammed the tailhook lever down and rolled my shoulders firmly forward in my straps to check my harness was securely locked. My first trap was sixty seconds away.

I was feeling slightly more relaxed with the flow of the pattern and the handling of my jet, but my work rate and situational awareness were at maximum. I hadn't forgotten that everyone there was trying to kill me. I reported abeam and confirmed my hook was down. The rest of the approach was another blur as I positioned my jet into the groove, called the ball and caught a wire.

I wasn't fully prepared for the violent deceleration of the landing. I was thrown forward in the cockpit, left hand fully extended on the throttles and at full power, as the arresting cable reduced my speed from 125 knots to 0 in about two seconds.

I had no time to congratulate myself on my baptism into naval aviation – there was work to do. I had to clear the landing area quickly to avoid the wrath of the boss. I pushed myself to remain focused and start the next phase of a carrier landing: taxiing around the flight deck.

The deck was ablaze with the coloured jerseys of the sailors and officers who worked in support of the carrier's air operations. Yellow Shirts, together with Green, Blue, Brown, Purple and White Shirts, all had a separate purpose in harmony across this potentially lethal environment.

They led me towards the catapult, steam hissing from its track, for my next dose of carrier aviation: the cat shot.

A Yellow Shirt guided me onto the catapult track. I glimpsed a T-2 climbing away from the cat shot, its exhaust plume blurring the multicoloured figures moving like ghosts in a desert mirage.

They swarmed around the track, preparing to hurl another jet, my jet, into the sky.

As I taxied, I reset my nose-up pitch trim to a pre-determined value and checked my fuel; this was critical in determining my weight, as the catapult steam pressure would be set accordingly. Another calculation I needed to get right – otherwise, I might get wet.

Once I'd confirmed the weight, the workers positioned me onto the catapult shuttle, attaching it with a wire bridle to two hooks under my fuselage. A Yellow Shirt signalled me to push my throttles to full power, and I was passed on to the waiting catapult officer in his White Shirt, fingers twirling above his head as he waited for my salute. My head went back as I snapped a salute, ready for launch.

I stared down the catapult track at the bow of the ship, and in my periphery saw the cat officer going through a safety check. Satisfied, he returned my salute and checked the bow was visually clear. In a choreographed movement, he lunged down on one knee as he touched the deck, pointing one arm forward with two fingers mimicking a pistol. This was the signal to the sailor in the starboard catwalk who controlled the catapult firing. He made one last safety check before he reached down and fired the catapult.

With a jolt, the cat accelerated my T-2 forward, slamming my head and body back into my ejection seat. I grunted. I was gauging the thrust and acceleration, confirming that this shot was 'good' and not 'soft', which might have thrown me into the ocean. In just over two seconds, my jet was accelerating along the linear track from a standing start to flying speed plus 15 knots.

Airborne, I whooped inside my oxygen mask – on the catapult no one can hear you scream, as we used to say.

What a ride. I was hooked. *Is Uncle Sam really paying me to do this?!*

'605, turn downwind. Hook down.'

'605.'

I had to maintain my focus until I completed my qualification. After all, everyone on the boat was trying to kill me. Three more sequences to go to finish my qual.

Safely back at NAS Beeville, I walked back to the hangar. I revelled in my sweat-soaked, smelly flight suit. I was walking tall and proud, filled with accomplishment and confidence. I was now a 'Tailhooker', the nickname for a carrier pilot.

It seemed like it was all over in thirty minutes, but my logbook shows the whole sequence was three hours and twelve minutes from launch to recovery.

# 11.

The time is 1257.

'Okay, guys,' I say, 'listen up.'

Any conversations are put on hold – it's time for me to brief the cockpit crew on the initial phase of my plan.

'We're diverting to Learmonth. We are flying, we have engines, and we'll keep flying. My plan is to enter a holding pattern over the airfield, and we'll continue to descend and prepare for our landing as we circle. We have flight-control problems, so we'll do a control check at 10,000 feet with the flaps extended. If the plane is okay, we'll fly the instrument approach for Runway 36. I will nominate instrument procedures for a manual landing. We're stuck with manual thrust and manual pitch trim, and I'm in manual control. They need thirty minutes to prepare for us, so let's take it slow and continue our prep. Questions?'

They solemnly shake their heads.

I pause, still commanding their attention. 'Looks like I picked a bad day to quit sniffing glue.'

We all chuckle at my reference to the classic film *Airplane!* –

or *Flying High* – a satire of the aviation disaster genre. For a very brief moment, our concerns are abated.

Our situation is bad; it's very bad. But it must be salvageable and the way out must be simple. That is my job and I must dig deep into my experience to pull a rabbit out of the hat. The plan better be a good one because I know things will happen quickly, like falling dominoes, down to our landing. I still have to work out how to negate any pitch-downs at low altitude. And once we land? Well, I won't be able to relax – we're in an isolated area, and I'll still be the captain until help arrives.

The time is 1300.

We are descending through 25,000 feet at our minimum speed and at a manageable rate of descent of 800 feet-per-minute. Our track is direct to overfly the runway. We can see it now, surrounded by the red-orange earth of Western Australia.

I hand over control of the aircraft to Pete; it's time for me to speak with my passengers and crew. 'Ladies and gentlemen, this is the captain speaking. Obviously, we are having some problems with our flight controls. Also, I'm aware that we have some serious injuries onboard, and we need to access medical treatment urgently. We are diverting to Learmonth Airfield, near Exmouth, and we should be on the ground in about fifteen minutes. Please remain seated with your seatbelts fastened, and we'll be on the ground as soon as possible. Thank you for your attention.' I'm trying to sound as calm as I can in the face of our situation.

The first phase of my strategy is complete, establishing ourselves overhead Learmonth airport. I am still pondering the scenario of a pitch-down at low altitude. How can I stop it?

I hear Pete commence broadcast calls to alert aircraft in the Learmonth area of our arrival. All aircraft airborne in our vicinity know we're tracking to Learmonth from our mayday call, but Learmonth is an uncontrolled airport, with no control tower or radar, and we are responsible for our own separation with other aircraft.

Pete says, 'All stations Learmonth, this is Qantas 72, an Airbus A330 inbound from the southwest on descent FL 200. Intentions are to hold to the south of Learmonth before landing Runway 36. Estimating circuit area time one zero. Qantas 72.' He has alerted all aircraft in the vicinity that we're operating as a commercial flight; our current altitude of 20,000 feet; and our landing intentions and arrival time in the airport area of 1310.

'I have control, Pete.'

I silently search my Filofax of military experience to find a strategy for addressing a low level pitch-down. If we pitch down again to the same degree, we could easily smash into the hard red earth short of the runway. I want to maintain speed for as long as possible before I slow down and prepare the aircraft for landing. I need options to cater for aggressive computer-commanded manoeuvres.

I conclude that I need energy at low altitude, and this energy will come in the form of speed. For a fighter pilot, speed is life.

Little did I realise I would face a one-versus-many battle in my commercial flying career. It's me against the confused and failing computers of this Airbus A330. It isn't a fair fight; knife fights never are. This fight is getting personal, and I will fight to the death.

*

We overfly the runway and start circling over Learmonth. I can adjust the diameter of the circle to maintain sight of the runway and allow time to descend. It also provides a reassuring image of the runway below to my worried passengers.

The easy part of my strategy is complete.

The next part of my strategy must now address the problem of erratic flight controls and the confused computers that manipulate them. The control check at 10,000 feet will confirm my level of controllability with the wing flaps extended. This is a common consideration in military flying, as the check confirms the aircraft's speed and manoeuvrability at approach speed. Checking the slow speed flight capability is best determined at a safe altitude before doing it for real.

I'm not sure how the computer's interaction with the erroneous data they've been sampling will affect our landing configuration and approach. It's an educated guess; all I know at this point is the computers are reacting to something I can't see and I don't trust anything they are doing. I need to ensure other aspects of control aren't affected by the multiple computer faults, and doing this at 10,000 feet will give me the space to deal with any more control difficulties.

The final part of my plan provides the tool for me to maintain energy and speed for the approach and landing.

As a student naval aviator, I was taught to fly the precautionary emergency landing (PEL) for each specific type of aircraft I flew. This was practised in the event of a real emergency involving issues with engine performance, degraded fuel systems or quantity, or a requirement to get a degraded aircraft on the ground quickly while still being able to use the ejection seat if necessary.

The PEL is a spiralling, descending approach. It starts at a medium altitude above the runway and corkscrews around in a constant turn to deliver you aligned with the runway centreline for a steep final approach to landing. Energy is maintained at a higher-than-normal speed to cater for an expeditious landing. I learned and taught this manoeuvre in both the T-2 and the Mirage, and even the space shuttle flew a similar pattern for recovery.

Flying a modified precautionary emergency landing procedure will give me energy in the form of speed at low altitude. I want to minimise the time in the landing configuration with landing gear and flaps extended because my manoeuvring potential is then limited. My thinking is to maintain a higher approach speed for as long as possible to provide energy to try to arrest, or restrict, any pitch-downs near the ground. My target is to be at 1000 feet in landing configuration and at the appropriate approach speed, a requirement mandated by Qantas policies.

It's the only strategy I can come up with that will give us a chance of surviving a computer pitch-down on approach. It's a gamble, but I'm managing the risk with the self-confidence to execute my plan successfully.

While I mentally assemble this landing strategy, I'm manually flying and navigating our possessed Airbus towards Learmonth, talking to air traffic control on my radio while monitoring Pete's call to the maintenance unit on the satphone. It's unlikely the engineer on the other end will be able to resolve our plight. I also have to keep Ross in the loop as he prepares the charts for our arrival. The near overdose of our body's fight-or-flight chemicals doesn't make this multi-tasking easy.

Imagine a juggler who balances rotating plates on sticks. At times, he has eight or more rotating at once on his table as he races from plate to plate, manipulating the sticks to keep the plates spinning near the horizontal as the sticks slow and the plates wobble. Ross, Pete and I each have our own set of plates that need attention.

As we enter the holding pattern and descend, I silently agonise over my strategy's effectiveness if I have to manoeuvre aggressively at low altitude. I know my other two pilots are also contemplating my actions if the plane does anything extreme at low altitude.

There are no published checklists or procedures to address our compromised situation, so I have to prepare for a last-ditch effort if things start failing close to the runway. Our jet is flying safely for the time being and I will brief them on the final bits of my plan very soon.

# 12.

The master caution continues to chime. *Ding.* What was that? *Ding.* What was that? At what point do I totally ignore this enormous distraction? Pete's finger stabs again at the emergency cancel button. No go. The chimes and dings keep sounding off, and the three of us glance at the centre display screen when they activate; we're hoping to see a message displayed long enough to allow us to interact with it.

Ross is on the interphone again with Lisa. He instructs her to remain seated with the belt fastened. She asks, 'Should I prepare the passengers and crew for an emergency evacuation, or will it be a normal landing?'

I overhear bits of this conversation, and I like the way she thinks.

Ross is looking at me, his eyebrows raised, for my answer.

I let slip a wry chuckle. 'Normal landing … as "normal" as it can be at this stage.'

'Normal landing at this stage,' Ross says.

I prepare myself mentally, and physically, for the next phase of our emergency management. Soon we'll come together and look over our plan, reading through the approach checklist.

I've been joking around with my two pilots a little, trying to help them to calm down and function better. My underlying message is: *Hey, if Kev is joking and relaxed, maybe 'The Fox' has it wired ... right?!* I need Ross and Pete performing and supporting me at the highest level for the next stage of our battle.

Pete is still on the satphone talking to the maintenance unit. But at the same time, he's making selections and setting flying targets while I change the aircraft's speed, altitude and heading. He's also monitoring the distance-measuring equipment, a ground-based navigational aid that provides distance information to our aircraft about its position relative to the airport; this is another manual tool we're using to help us maintain our situational awareness. In between these tasks, Pete continues to broadcast information to other aircraft in the area as we descend and establish ourselves in the holding pattern.

'All stations Learmonth. Qantas 72 entering the visual hold overhead. Descending to One-Zero Thousand. Qantas 72.'

I look over at Pete and give the finger-across-the-throat signal: time to end the call. Maintenance are unable to resolve the problem, and it's causing further distraction. We can't afford any more distractions right now.

*

The time is 1315.

For the past ten minutes, we've been slowly and cautiously descending in our circular pattern over the airport. We're approaching 20,000 feet, and now we need to come together for the next, and most important, phase of our emergency.

We have to prepare to get this baby on the ground.

Pete glances over at me. 'I can't leave you guys alone for one second,' he jokes.

Ross and I chuckle as I shrug. Pete's quip is a good sign; he's actively managing his startle response, too, and I know his awareness is where it should be.

'Let's have a good look at the status page now. Ross, prepare a brief for the approach to Runway 36.'

It's time to fine-tune our approach and landing plan.

Pete starts reciting the litany of inoperative systems. How will they affect our landing? There are some big-ticket items listed, each one represented by an abbreviation that I instantly recognise but would confuse any layperson. It is an overwhelming list and we examine it carefully. Lots of systems aren't working but I fall back on the things that are: the engines, the landing gear (which can be extended manually) and, of course, the three pilots.

First, we must use a flap setting less than the normal full landing flap. Our approach speed will be faster than normal, which is fine with me. It will affect our landing distance calculation; fortunately, the runway is 10,000 feet long.

Second, the automatic braking system isn't available. Normally, the brakes will activate automatically as the aircraft makes contact with the runway, but I'll have to apply them manually. I glance at our gross weight: 183,500 kilograms. We're 3000 kilograms less than maximum landing weight, but I must be gentle on the brake application to prevent overheating. Too much temperature on the brakes transmitted to the tyres could cause them to deflate or explode. This must also be factored into our landing distance calculation.

Third, some of our spoilers aren't available. These are panels on top of the wing that deploy on landing, 'spoiling' the lifting force of the wing and putting the weight of the aircraft onto the landing gear to optimise braking. Our landing distance calculation must also factor in the loss of the spoilers.

Finally, the automatic thrust system is off. I had deselected it earlier in accordance with the unreliable speed condition. Normally, the automatic thrust system is linked to the sidestick movement during a manual approach; when you pull back on the sidestick, power is increased, and vice versa to reduce power. This isn't a significant item for me to manage, because I practise manual thrust landings routinely, but it does add another factor for the landing distance calculation.

'Okay, Pete. We'll use the GNSS Approach RW36.'

A GNSS (global navigation satellite system) approach is a non-precision approach using GPS satellites along with latitude and longitude waypoints, instead of the raw navigational beacon points from land-based aids. It can position us more accurately with the runway to aid in my visual transition to landing.

Pete nods as he dives into his navigation keyboard to select and load this approach into our computer flight plan. Then I hear him quietly muttering in frustration as he tries to interact with the computer. 'It won't load, Kev.'

I take it in stride and instinctively improvise. 'Okay, visual approach, 10-mile final, using visual procedures. You can back me up with the distance and recommended altitudes on final.' Each approach chart has an altitude-versus-distance scale to assist the pilot in managing the correct vertical profile to the runway and we'll just use our eyeballs to help us get to where we want to be.

I've nominated visual approach and visual procedures because these will keep our procedures simple. All pilots will be 'heads up', looking outside as we progress on the approach. Pete can access the raw distance transmitted by the airport's navigational beacon and compare it with recommended altitudes to confirm I'm flying a three-degree approach path.

'Ross, check landing distance with flaps 3, two spoiler pairs inoperative, manual brakes, wind calm and landing weight 184,000 kilograms.' I know we can land comfortably with the runway available, but this is a required computation, especially for our degraded landing configuration.

Ross inserts these parameters into our 'performance' laptop that is pre-loaded with take-off and landing software and will assist in the calculation. It's an industrial strength clunker of a laptop and it weighs a tonne. He hands the laptop to Pete for a second independent calculation and confirmation. Pete gets the same figure as Ross, meaning we have enough runway to land safely. It's like a gameshow without a prize.

The pace of preparation is increasing and no time for my characteristic flippancy. Its absence shows I have my game face on.

Although our nosewheel steering system isn't on the list of failures, I note that I'll check it on landing and mention this to Pete and Ross. It's also computer controlled and, if it's affected, I won't be able to turn the aircraft after landing.

Pete and I discuss the unusable airspeed information on my primary display; it's still jumping around erratically. Normally, the flap limit speeds and the computed approach speed are nicely displayed as coloured markers adjacent to the numerical speeds on the vertical speed tape display. I won't have these visible cues available. Pete confirms he'll manually set the speeds as he has continued to do while we extend the flaps and slow to final approach speed. I must keep crosschecking my speed performance with the standby instruments and Pete's primary display, as I've been doing since my display became unusable.

I've been carefully stacking the dominoes; our preparation is almost complete. Now I must formally verbalise my plan and discuss our limitations and contingencies. The cockpit voice recorder will capture everything I say.

'Ready to brief? Okay, we have multiple inoperative systems, which we have reviewed. We have plenty of fuel for multiple approaches and we are below maximum landing weight. We'll do a control check around 10,000 feet. If that's all okay, we will descend and shoot for a 10-mile final at 3000 feet. I'm going to fly a high-speed approach, and I hope that the higher speed will allow me to try and arrest any nose-down manoeuvres by the computers at lower altitude. I'll use the rudders if I have to and minimise altitude loss.'

These comments are met with solemn silence; I'm briefing them on my last-ditch manoeuvre to stop us from going into the ground at the hands of our confused computers. I'll use full rudder input if necessary to keep the nose above the horizon should the computer try to execute another pitch-down.

'It will be visual procedures for a visual approach. Manually flying. Manual thrust. Manual pitch trim. Manual braking. Manual Fucking Everything,' I add in disgust.

'I'll use the full length of the runaway to keep the brakes cool, and I'll check the nosewheel steering on the runway. If we have to go around, I'll clean up and fly downwind for a visual circuit. Any questions?'

They nod their heads in understanding.

We're now in the serious phase of flight. It's time to land.

'Approach checklist, Pete.'

Pete runs through a challenge-and-response checklist to confirm we've covered all the required items to commence an approach to landing. The issue with PRIM 3 is presented again in our review of inoperative systems, and I reluctantly agree to shut it down. Turning it off is different to resetting it for operation. Fortunately, we remain flying.

'Approach checklist complete,' Pete announces.

'Okay, let's have a look at the runway and parking apron.'

I manoeuvre our jet to a point abeam the runway and dip our left wing down. Below us is Learmonth Airfield. We crane our necks to see the runway markings and turn-off points from the cockpit windows on the left side. One-third of the way down the runway is the turning area to access the parking apron; we'll have to stop on the runway, turn around and taxi back along the runway to park in this area. We see the airport terminal and another support building; it's a far cry from the international airports we are used to servicing. There are no aerobridges here, so we'll be parking somewhere on the tarmac and use external steps. Some support vehicles are parked nearby.

I extend upwind as I continue to descend.

Already, I'm preparing myself for the landing. I know what my body will try to do to me when we're safely on the runway; it will flood me with all the chemicals that make up the 'post-adrenalin-you-are-safe' phase, to assist the body in purging the adrenalin released during our death dives. Landing safely is only one phase completed. We have a long day ahead, and I can't allow myself to come down until all my passengers and crew are safely in Perth.

# 13.

Our altitude is 11,000 feet, and I reduce the power and increase the nose attitude to stop our rate of descent, feeding in manual nose-up trim through the big black trim wheel next to my knee as we decelerate.

'Control check,' I announce.

We are getting down to business now and entering a phase of uncertainty. The flaps will be extended to ensure that they will actually work and that the jet will not react adversely. The three pilots are all working very well together and our minimal communication is strictly procedural. I know everyone in the plane is thinking the same thing that I am … will we land without any further drama? The 10,000-foot runway below is a big carrot dangling in front of us and we must be patient to ensure we can get there safely.

It's my responsibility to ensure we can and I am prepared.

We're at our minimum clean speed and below the maximum speed for the first phase of flap extension. I continue to manually trim the nose up as the speed decreases.

I order flap extension to Position 1.

Pete reaches down to move the flap handle.

'Speed check. Flaps 1.'

Pete repeats my request, checks the speed is correct for extension, and selects the flap lever to the first detent, Position 1. We watch the progress of the flap extension on the upper screen, which shows its movement in green as my grip on the sidestick firms. It stops at Position 1, and Pete winds the speed back through the selection knob. 'Flap 2.' Pete repeats the process and selects the flap handle to the second detent. He sets the speed to approximately 180 knots.

So far, so good. The flaps have extended normally, and I complete an S-turn to the left and then to the right, checking for any abnormal feel or stick response.

Good enough for me! I'll land in this configuration if I have to.

'You guys happy?' I check. I'm really asking if they're satisfied that the aircraft is controllable and that we can continue our approach. They nod and flash a thumbs-up as I push the nose down to commence our approach.

'Set Altitude 3000.' This is my altitude target to intercept the 10-mile point on the centreline from the runway. Pete dials in 3000 feet through the altitude selector knob. The yellow command bars on my primary display are giving me good targets to fly, but I keep crosschecking my attitude and speed with the other instruments.

Pete selects the PA button on his radio control panel and announces to the passengers that we've started our approach – and that they must remain seated with belts fastened. We keep them informed now as we get closer to landing.

The mood on the flight deck is tense as we descend. Three pairs of eyes are scanning the instruments, watching my manipulation, ready to announce any deviations from a safe flight path. The stall warnings and the master caution *dings* continue. I know I'm not stalling but it's very spooky to be descending towards the runway with these serious warnings ringing in our ears. These audible distractions add urgency to our situation; the last thing any pilot wants to hear is the stall warning as the ground gets closer.

The time is 1325 and I'm descending rapidly, maintaining 180 knots. We're approaching 5000 feet of altitude, heading downwind away from the airport. I am positioned to turn left towards the airport now. I want to be at 3000 feet of altitude, 10 miles from the runway. The higher speed and minimum power are making the manual flying a bit easier. As we carve off altitude, I take a few deep breaths.

It's show time. The dominoes are falling fast and furious as we approach the runway.

I'm on full alert now. My shoulders crunch forward; my instrument and visual scan rate are at 150 per cent. I'm tapping into my body's chemicals again to focus on delicate movements of the control stick, adjusting the trim wheel as we decelerate, while smoothly manipulating the power.

I'm back in the physical and mental space of a carrier landing, but I'm feeling very vulnerable. The pilots are reaching a critical point in our emergency. The lower we get, the less altitude we have to recover from computer-commanded manoeuvres.

The tension is building as we approach Runway 36.

*

The time is 1327.

Pete makes his final broadcast to traffic in the Learmonth area. 'All stations Learmonth. Qantas 72, an IFR Airbus A330. Ten-mile finals. Runway 36.'

Then he makes a another public address.

*

In the cabin, everyone listens closely to Pete's announcement that we're on final approach. He also instructs passengers to follow crew instructions.

The passengers and cabin crew are already aware that our landing is imminent – they can clearly see the burnt-orange earth surrounding the airport.

The cabin is very quiet. Some people are still crying, but everyone on the flight is hoping, praying, for a safe landing.

The end of their nightmare is a few minutes away.

*

'Speed.'

We are closing in quickly on the welcoming runway. Pete is doing his job, reminding me that my speed is higher than it usually should be. But this isn't a normal approach or circumstance.

'Acknowledged,' I reply.

Speed is my friend now. My whole body is on alert for any external control inputs that I'm not making. I am still flying through a computer and the warnings are continuing. I don't trust anything the computers are trying to tell me now. If the plane starts to pitch down, I hope my high speed will give me a chance to snatch the nose up before I am denied control again. It's a big weight on my shoulders. Our approach is like running through a hidden minefield.

I'll work towards slowing down soon, but not quite yet.

'Gear down,' I announce to Pete, as I point at the lever and gesture in a downward motion. The extra drag of the landing gear will keep my speed under control and help with my descent.

He pulls down the landing-gear handle.

The gear doors snap open and the aircraft vibrates as the gear extends. When the gear locks into position, the noise level increases in the flight deck. We watch as the upper display screen confirms gear down and locked.

'Gear down,' Pete confirms. He reaches across the centre instrument console and moves the speed-brake lever up, arming our working spoilers for deployment on landing. They will automatically deploy when the main wheels touch the runway and assist with the braking.

I check my rate of descent against my intercept point. I'm a little high but I can accept that. My energy is good; speed is life.

I take my eyes off my instruments for a quick look up towards the runway. The lights are all on, thanks to Ross, and I confirm that I'm slightly high. I take another deep breath as I align QPA with the runway.

As we descend towards 2000 feet, my anticipation grows. The stall warnings and chimes continue; they're more significant at this low altitude. All three pilots scan the centre display screen and monitor our performance.

I glance at Pete. 'Are you satisfied that we aren't stalling?'

We must both confirm that my energy and flight path is safe because the aircraft is warning us that it is not. No confusion regarding our aircraft's safety can exist. With these warnings still active, the plane could conceivably pitch down again.

'Yeah, Kev,' he replies. 'I'm happy.'

Our capability to recover from any computer-generated upsets is diminishing as our altitude decreases and we approach the runway. I'm still locked on to my instrumentation, searching for errors or faults while confirming that my approach is stabilised for landing.

The three pilots are working together, each physically and mentally coiled like a rattlesnake with anticipation. In my mind, I recite the Top Gun prayer: 'Dear Lord. Please don't let me screw up!'

'Flap 3. Final flap. Landing checklist,' I announce as I level the aircraft at 1500 feet.

Pete selects the flap lever to Position 3 and recites the challenge-and-response landing checklist: 'Auto thrust.'

'Off.' It's a manual thrust approach.

'ECAM memo.'

'Landing. No blue.'

This is an automated systems check to ensure the aircraft is configured correctly for landing. 'No blue' means there are no unactioned items affecting our landing that are recognised by the automation or displayed on our centre screen. At least this system is working.

Both pilots confirm: gear is down, automatic spoilers are armed, flaps are in Position 3 and seatbelt signs are on. The wheel system is displayed on our lower centre screen and also confirms our landing gear is down and locked.

'Speed,' Pete prompts.

'Correcting.'

I'm gradually increasing the nose attitude as the aircraft decelerates to our planned final approach speed. I advance the power levers and adjust the trim wheel to match the speed. Normally, thrust control and pitch trim are automated functions – but not today.

We're nearing the runway; it's about 1000 feet below us. I continually probe and confirm, through my sidestick inputs, that I'm still in control of the jet. The glidepath lights show I'm maintaining the correct rate of descent and aiming to land 1000 feet in from the end of the runway. Two big white stripes painted on the runway mark my aiming point.

The approach is stabilised, and the runway is growing in our windscreens. I'm running out of altitude should anything adverse happen, but I stay alert and continue with the task of flying the plane. I adjust the position of my feet on the rudder pedals, just in case I need them.

'Manual braking. Go around altitude 3000.' This is my verbal reminder to the pilots that our automated braking system is also inoperative; we'll climb to 3000 feet if we have to discontinue this approach.

'500 feet. Stable.' Pete confirms we're 500 feet above the runway, and confirms speed, glidepath and rate of descent satisfy the 'stable approach' criterion.

'Roger.' I commence scanning up the length of the runway, using my peripheral vision to gauge the point I will raise the nose and land. I mentally pinch myself, a reminder to 'stay up', and to prevent my body from flooding my system with 'You're safe now, Kev' chemicals. The end of this flight is still a long way off.

In my peripheral vision I see the 'piano-key' stripes that mark the beginning of Runway 36. They're passing underneath the aircraft as I raise the nose to reduce the rate of descent for landing. Before the wheels contact the runway, once again I remind myself to 'stay up'.

*Thud.*

I smoothly reduce the thrust to idle as the main wheels touch down, then I pull the engines into reverse thrust and vector their exhaust forward to help us decelerate.

As the nosewheel makes contact with the runway, we can hear wild cheers and applause through the dense metal of the bulletproof cockpit door. We streak past the turn-off for the parking apron. I quickly glance to the right and catch a glimpse of the emergency response waiting for us: two ambulances and a volunteer fire truck, side by side near the parking area. I have to turn around on the runway and taxi back to park there.

I maintain the engines in reverse and gradually apply the brakes as the aircraft decelerates through 100 knots and down to taxi speed. I know I have to be gentle as my feet press forward on the rudder pedals to initiate the manual brakes, and I'm careful not to push too hard: hot brakes would further complicate our situation.

'Seventy knots,' Pete announces. This is his prompt for me to deselect reverse thrust in order to prevent the engines stalling at slow speed.

'Roger,' I say as I stow the engine reverser levers to the forward position.

I glance at my two pilots, their faces awash with relief that we've safely landed our stricken jet. 'So, a little excitement in an otherwise dull day,' I quote, in my best Arnold Schwarzenegger accent from the movie *True Lies*.

They look at me with their mouths slightly open in astonishment, before giving me exhausted and satisfied grins.

I'm anything but relieved. I know we've cheated death, and no encounter with death is ever fulfilling.

# 14.

Qantas Flight 72 has landed safely.

In the cabin, crew members rush to access first-aid kits. Despite their injuries, they've found the courage and energy to attend to the injured throughout the aircraft. But not Fuzzy. He remains strapped into his seat, seriously injured.

I take a deep breath and mentally open a new procedural box: we must now taxi safely to park our stricken Airbus.

As we decelerate towards taxi speed, my left hand slides down to the nosewheel-steering control in front of the control stick. I move it left, then right. The nosewheel deflects accordingly, the aircraft moving as I command. I turn our big aircraft around and go back down the runway to reach the parking area.

Even though we're on the ground, the pilots must maintain an operational discipline. The unfamiliarity of Learmonth Airfield keeps us on edge as we taxi towards the small terminal. I know that as soon as we shut down our engines, we'll be inundated with calls from headquarters seeking information. I'm not looking forward to this.

As we enter the apron we see support staff from the airport, signalling for us to align the aircraft with a yellow line painted on the tarmac. We are the only aircraft at the airfield, but all of us visually check our obstruction and wing clearance as I turn the A330 to align with our parking position.

I select the parking brake; Pete deals with switching our power source; and Ross orders the cabin crew to disarm our emergency escape doors. I hear the *thunk-thunk-thunk* as crew members slide the disarming levers on the two front doors.

I switch off and secure our engines. As they wind down, for a moment we can stop and reflect. I hold out my open hand and the three of us clasp hands like The Three Musketeers for a job well done. I know this event will bind us together for the rest of our lives.

But all is not at peace – the flight deck still resounds with abbreviated *ST-STA-ALLL* warnings and the incessant dinging of the master caution messages. We are dumbfounded and disgusted. The computers think we are still flying and continue to warn us. My hunch that they were all false is vindicated.

The time is 1342. The Qantas Flight 72 mayday and diversion are complete.

'Pete, shutdown checklist. Ross, you can open the door.'

As they carry out these duties, I reach down to get my mobile phone out of my bag. My daughter is on holiday in Europe with her boyfriend, but I'm sure our event will be broadcast around the world. Contacting her is my highest priority; she'll work out very quickly that it's my flight plastered on CNN.

I send her a short, coded text message: 'I'm okay and I love you.'

The text is sent just as my phone starts ringing. I don't recognise the number but, obviously, it's someone who knows me.

'Sully, it's Scoob. Are you okay, mate?'

It's Captain Steve Anderson. He's the welfare officer from our pilots' association.

'How'd you know to call now? I'm a bit busy.'

'We've been tracking you! What happened? Are you injured?'

*What happened?* A question I still have no idea how to answer.

'I'm okay,' I say, 'but we have severe injuries. I don't know what happened. She was a very sick jet and she wouldn't let me help her. I gotta go.'

As I finish the call, Lisa wanders into the cockpit, her arms limp at her sides like those of a zombie. She's distraught and shell-shocked. 'What the fuck was that?!'

I take her hand and squeeze it. I know it must have been hell in a cabin full of frightened and injured people. No airline training would have prepared her – or any of the crew – to deal with this scale of casualties and violent manoeuvring. The good news is that she isn't physically injured, thanks to Pete's heroic actions.

'I don't know,' I tell her, 'but we're safe now. I need you to coordinate with the medical staff when they come onboard. They're moving up the steps to the aircraft now. Do you need anything? Are you okay to do this?'

I keep holding her hand as she nods. Our hands separate, and she disappears back into the cabin. Moments later, I hear an entry door open; the emergency responders enter. Pete makes another announcement, telling passengers to remain near their seats and to keep the aisles clear for arriving medical staff.

The satphone is ringing now. It will be the company calling from the Integrated Operations Centre (IOC) at our airline headquarters in Sydney. The master caution dings and stall warnings continue, adding confusing background noise to our discussions. Before I sign off, I explain that we have two live dogs in our cargo compartment and request assistance to unload them before it gets too hot.

Hopefully, they can arrange a cargo technician to come up to Learmonth with everyone else. We say our goodbyes.

*

With a break in the action, Pete is accessing the post-flight report from the aircraft's maintenance computer. Any system failure or fault is recorded and can be accessed readily after the flight. Engineering and maintenance personnel routinely use this summary to rectify any defects before the next flight.

'Let's have a look to see what happened,' Pete offers, with the air of a TV crime-scene investigator.

The paper summary streams out from the slot of our small onboard printer, the inkjet rattling and whining with each row. Pete reads it like an old stock market tickertape as he feeds it through his hands.

'Did you order a copy of *War and Peace*, Kev?' he asks, as the printer continues to spew forth the report.

'Ho-*ly* shit. Check this out!'

We huddle together and examine the printout. At time 1240, just minutes after I returned for duty, ten simultaneous failures were recorded.

The first group of failures originated in the internal reference and global positioning module that form part of ADIRU 1. Only one of them was presented to me and Ross for action – a fault in the navigation part of that unit. But now we can see the cascading effect this fault inflicted on the many other systems that are dependent on position data from this computer.

Other faults were recorded to systems such as the captain's navigation map display and the two channels of the global positioning system. Critically, faults in the electronic flight-control system were recorded but weren't identified to the pilots. These are the computerised systems that maintain the aircraft within the safety of its defined operational boundaries.

Additionally, a fault in the air-data module of ADIRU 1 was recorded. This is the internal module that converts and distributes air data, such as airspeed, altitude and angle of attack, to appear on the pilot's primary flight display and for use by the PRIMs. Why wasn't this fault revealed to us? If it had, the flight would have been completely different. That portion of ADIRU 1 would have been turned off and we would have proceeded to Perth without injuries, loss of control or unreliable speed.

Over the next five minutes, another barrage of failures was recorded: the autopilot disconnections, faults to the primary flight-control computers (PRIM 1 and PRIM 2) that weren't displayed, and a fault to PRIM 3 that was displayed. The change to the Alternate Law system of control was recorded at 1245.

We sit in shocked silence. Now it becomes clear why we unable to comprehend what was happening and why our Airbus went berserk: these are serious failures that removed the automation we were trained to rely upon. The computer systems became confused and continued to warn us of serious dangers that weren't present. This gave us many opportunities to get it wrong – the path to potential pilot error is watermarked all over this summary.

I make a quick assessment as I stare at the summary. We had survived a total meltdown of our electronic flight-control system, along with an overload in the operation of the centralised maintenance control computer. This prevented us from interaction and reconfiguration to recover essential systems such as automatic pilots, trim and thrust, pressurisation, braking, and accurate airspeed and altitude information.

This summary doesn't fully explain what exactly happened, though. Were protection modes activated to save us from what the computers perceived as an external threat? If so, which modes and why? This critical information is not displayed to the pilots nor is it part of the Airbus design.

'I think we are pretty fucking lucky to be sitting here,' I conclude.

Ross and Pete nod solemnly and their eyes drop in comprehension.

This wasn't supposed to happen, but it did, and now we're left to clean up the mess.

We've been parked now for about fifteen minutes.

Pete gets out of his seat. He needs some space to ponder what could have been. 'I'm going out into the cabin and do a circuit, show a face,' he says. At my nod, he ducks out through the open flight-deck door.

With pen in hand, I retrieve the summary and flip it over to the blank side. 'Okay, Ross. Let's make our own summary of what happened while it's fresh in our minds.'

Through the post-adrenalin haze, we start re-creating the sequence of events, using the post-flight report's time markers as reference points. While we try to concentrate on this task, the flight deck is still dinging from the system faults, and we continue to get the truncated *st-a-all* warnings. Unbelievable. The aircraft's computers still think we are airborne!

The satphone is ringing again. It's a representative from our engineering department who wants to give us some guidance in preserving the flight data and cockpit voice recorder: the aircraft's black boxes (they're actually orange). We copy down the procedure, then action the appropriate switches on the overhead instrument panel to prevent this data from accidently being erased.

The engineer offers to help us de-energise the flight warning system. *Finally*, we'll have some quiet on the flight deck. The incessant warnings are a constant reminder of our fucked-up flight.

Ross and I copy down the selection of circuit breakers that will be pulled to de-energise the computers of the flight warning system. They live in the electronics bay, under the flight-deck floor, and I vacate my seat to access the trapdoor and ladder for Ross. He disappears down the ladder. Within a minute, the alerts have been disabled as the warning systems computer is deactivated.

Silence at last!

Ross climbs back into the flight deck, and we secure the small access door. We continue to work on our summary in Pete's absence.

*

I'll learn later that the QF72 mayday call at 1254 prompted an emergency response that mobilised the assets of the entire state of Western Australia.

The WA Police activated their Emergency Operations Centre to coordinate the rescue operation at their headquarters in Perth. This facility was commissioned only a few months earlier, and the QF72 emergency would be the first test of the centre's capability.

Ten minutes after we declared the mayday, the Centre received notification of a Qantas Airbus diverting to Learmonth with instrument failure. Twenty minutes later, the flight number was confirmed, along with the fact that we would be making an emergency landing.

The Exmouth regional medical centre was informed of the requirement for medical support. This centre mobilised its two ambulances, a doctor and two nurses to meet the aircraft. Concurrently, the offices of customs, immigration and quarantine were notified of the emergency diversion; they scrambled to prepare for travel to Learmonth.

When QF72 landed safely, the Australian Transport Safety Bureau was notified of the situation.

Within minutes of the aircraft parking, the first casualty reports started to filter in. They identified fifteen casualties involving severe lacerations and fractured limbs. The Emergency Operations Centre coordinated with the Royal Flying Doctor Service and mobilised their aircraft to provide medical support.

The Centre also organised charter flights to airlift police officers from surrounding areas into Learmonth to assist with the emergency response.

At 1408, medical staff from nearby Carnarvon Hospital were mobilised to assist. Over the next hour, the numbers of injured were updated and serious injuries were reported. The State Crisis Centre was activated at 1540.

We soon witness the amazing effort of the WA emergency response.

# 15.

Pete returns to the cockpit from his inspection of the cabin, looking visibly affected by what he just witnessed. He is distracted and deep in thought as he takes his seat.

I look at Pete. 'Bad?'

He nods and slumps back into his seat. His face is flushed and his injured nose is showing swelling.

'Keep an ear to the radios, Pete.'

I've been waiting for him to return. It's now my time to 'walk the walk'.

The scope of my next course of action isn't lost on me. I'm still the captain, and there are no company representatives present to relieve me of that responsibility.

I've been trained in the responsibilities of 'keeping watch', an integral requirement for every navy man. A ship at sea never sleeps; it operates 24/7. In times of reduced activity, watches are set. Every member of the ship is rostered to keep watch, even the pilots in the Air Wing, and there are specific duties every watch-keeper must follow. The eleven General Orders are taught to every member of the Armed Forces in basic training, and committed to memory.

The Fourth General Order states: 'To quit my post only when properly relieved.'

As the captain, QF72 is my watch. I will remain on duty until relieved – whenever that will be.

Leaving the flight deck, I come across a group of medical responders. 'How's it going?' I ask.

I know they're overwhelmed by the number of injured passengers and crew; it's etched on their faces. They tell me they're progressing towards the rear of the aircraft and treating the more badly injured in that area, but they've run out of neck braces.

'How many did you bring?'

'Twenty-six, our entire stock at the medical centre.'

*Shit.*

'Okay. Are you able to source more? Can we assist through our company? Will you be looking to disembark those who are able to walk soon?'

They're working on it, they say, and I leave them to continue with their duties. I won't micro-manage them but I will certainly support their efforts in any way I can.

I've decided that my first public address to the cabin won't be made from the cockpit. My passengers and crew will be looking to me for leadership and compassion, so they need to see me as I talk.

From the entrance to the cabin, I can see the many seat rows and people standing in the aisles. It's my first look at the destruction caused by the pitch-downs. I pull the cord on the communication handset to its maximum length and start my announcement. 'Ladies and gentlemen, this is the captain.' I hold up my hand so all can see me in the cabin, even those still seated.

Heads jut out from every seat row as far back as I can see.

'Welcome to Learmonth.'

This is met with cheers and applause from everyone who is able to celebrate.

'We are safe now, and we are working to get you off the plane as soon as we can. We're communicating with our company, and they're working on a rescue plan to fly us out of here and down to Perth. I thank you for your patience and understanding. Please remain near your seats as the medical team assists those who are injured. I will be making my way through the cabin now. Thank you for your attention.'

Stowing the handset, I start a walk that will change my life.

The cabin looks like the aftermath of two opposing armies engaged in hand-to-hand combat on the Western Front. There's blood on the walls, and on the faces and clothes of the injured. There are hideous star-shaped holes in the ceiling that look like the Incredible Hulk punched them. They hold the remnants of the hair and bloodied scalps of those whose heads have punctured the aviation-grade plastic. The resulting wounds could easily have been made with a large serrated knife, raked along the victims' scalps.

Bloodied bandages and pillows are strewn in the seat rows and aisles, mixed with the debris from broken luggage lockers and ceiling panels.

Can you actually see a person's heart as it breaks? Parts of my psyche are being brutally ripped out of me as I move through the war zone that was once the cabin of an A330. I try to remain strong. But as I speak with my passengers and my courageous crew, seeing their injuries, feeling their grief, my heart is cracking.

Moving down the aisle, I'm asked the same question by my bewildered and bruised passengers: 'What happened?'

What can I say to them? I answer sincerely that I don't know exactly what happened but I tried to stop it. It's a bad state of affairs when the captain of his aircraft doesn't know the reasons for its violent behaviour.

We have many young children onboard, some secured in bassinets on the bulkheads and others on their parent's laps. I'm seeing them now as I move into the economy section. The parents are holding their children and trying to console their pain and tears. All of them have been injured in some way; most likely they were playing on their parents' laps after finishing lunch. Contusions are developing on some, while others are wearing bloodstained bandages.

My heart is dissolving; as a parent, I sympathise with those who've helplessly witnessed their children being injured. Their accusing, silent glares burn into me like lasers, communicating the unsaid message: 'Look what you did to my child.' These images are seared into my memory. I can only offer sympathy and support as I continue my torturous walk towards the rear of the cabin.

My best poker face, which I wore when I entered the front of the cabin, has completely dissolved by now. But it isn't until I enter the rear galley that my heart receives a death blow.

The floor is littered with broken shards from the economy-class plastic cups, as if sprinkled with multicoloured icicles. Beside them lies Fuzzy, legs immobilised and splinted. I kneel next to him and clasp his hand in a brotherly grip.

What can I possibly say to him? My face must say something that no words can convey. He thanks me and thanks God that I'm his captain.

Tears are welling in my eyes as I try to comfort him. My face is burning in anger, in frustration, in compassion, as I release his hand. My heart splinters.

The stainless-steel handhold connected to the galley bench has been bent … Did a metal meal cart hit that bar, or a human body? I shudder in disbelief at the force required to deform that bar.

I must keep moving; I must continue my cabin inspection.

Across from Fuzzy, still in the crew seat, is Peter Casey. He has a bandage on his forehead partially covering his badly bruised left eye; blood is seeping through. His arm is immobilised in a sling, and his shirt is covered in a mix of blood and red wine. He looks like the victim of a violent home invasion, and I gaze upon him in shock and sympathy.

Peter and I engage in some brief conversation; as a pilot, he's curious to know the reason the plane manoeuvred so violently.

'Oh my god, Peter. Are you okay? Is that blood on your shirt?'

'Some of it is, some is from the red we were drinking before we got knocked out. What the hell happened?'

I repeat my feeble answer that the computers went berserk, adding that I still don't fully know the reasons but they weren't pilot-induced or turbulence-related. My face is flushed and my eyes are wet with suppressed tears as I leave Peter in the care of the medical responders.

Slowly I make my way forward again. I can't stop staring at the cabin's ceiling, and my memory permanently records the busted plastic holes still holding on to hair and bloodied remnants of scalp, the broken luggage lockers, the dislodged ceiling panels and exposed ducting. It's a nightmare reality, not of my doing, and not in my power to prevent.

It's very lucky no one lost their life during the pitch-downs.

A male passenger asks me to explain the aircraft's behaviour. I offer my canned reply, which he accepts. I must look sad because he asks: 'How are you doing?'

'I'm okay, thanks. It's just going to be a long day.' I'm a lousy liar.

Passing the forward galley again, I meet with some of my crew. Samantha has a bruised eye, Kimberly can't move her neck without pivoting from the waist like a robot, and Tasha is favouring an injured shoulder. Rory is active in his zone, interacting with the passengers there.

This is such bullshit … how could this happen?

I sense that throughout the aircraft, the uninjured passengers are becoming impatient with having to remain onboard; we need to organise their disembarkation as soon as possible. I return to the cockpit, and Pete volunteers to pursue it with airport staff out on the tarmac.

Back in my seat, I look for my uniform tie in the stowage area near my side window but I can't find it. Then I check my mobile phone, hoping for a message from my daughter, and discover the reception has vanished.

*

Pete liaises with the local airport staff to disembark passengers and move them into the terminal. Although it's spartan and not designed to handle this many people, the local staff are bending over backwards to make it work.

It takes about forty-five minutes for the plane to be nearly emptied; those who remain will have to be stretchered off due to the serious nature of their injuries. I stand next to the forward door, offering encouragement and asking again for patience as the able-bodied passengers file past. I see relief, worry and pain on all the passengers' faces, but I also see gratitude sparkling in their eyes.

I move through the aircraft again to see who's still onboard. Pete is talking with Diana Casey, offering her a drink of water. One of my crew keeps an elderly couple occupied with conversation. Another passenger is still immobilised on the floor with medical staff hovering around. I glimpse Rory sitting silently with Jill. They appear comatose but I know that the shock of our ordeal is starting to sink in.

In the forward galley I grab a soft drink and a packet of nuts, wanting to keep my sugar levels high to help me cope with the hours ahead.

As the plane empties, my cabin crew are assembled in the business-class area, remaining clear of the aisles. Of the nine crew, only two aren't physically injured. Fuzzy is still in the rear galley waiting to be stretchered off the aircraft. All are shell-shocked, traumatised, their eyes open in a blank, thousand-mile stare; they've endured a hellish, life-threatening event. It's my duty to try and explain it to them.

I'm mentally preparing to deliver an operational debrief, which I'm required to give to my crew. I wish I knew more about the causes of our emergency, but this is more an opportunity to check on their welfare and answer their questions. It's also a confidential briefing, so I wait until we have a lull in the cabin activity before I huddle them around me.

I've never sugar-coated any explanations of technical issues to my passengers or crew; I explain everything in a way they'll understand, without scaring them, and answer their questions as best as I can. This debrief is no exception. I take the time to describe the problems Ross and I faced during and after the pitch-downs.

'We basically got ambushed. I didn't move the aircraft and it certainly wasn't turbulence. It could only be the computers.'

'Can't you override it?' Kimberly asks.

'No. It's not part of the design.' I answer bluntly.

I also explain that the computers acted without warning, and this scenario of computer-generated manoeuvres in reaction to false data isn't covered in our Airbus manuals. I discuss our diversion to Learmonth as the best choice, based on the serious injuries in the cabin and our distrust of the flight-control systems and the computers.

'I'm sorry I couldn't let you out of your seats. I wasn't sure if the computers would do it again and I had to keep you all safe.'

They listen intently and shed a few tears, not only from the pain of their injuries but also in the confirmation that they've faced a near-death event.

'I need a drink!' a crew member confesses as I finish the debrief.

Using my command prerogative, I 'order' someone to find some champagne and glasses. My crew deserve a drink for their heroic actions, and it's a worthwhile distraction. They rummage through the forward galley lockers in search of champagne bottles. Fortunately, some have survived, but only plastic cups are available. The corks are popped, the cups filled.

I salute my cabin crew and offer an old pirate toast. 'Let's have a drink while we still have breath, for there's no drinking after death!'

I take a sip but it tastes like I'm sucking on an old one-cent coin. It's a strange reaction from my palate and my body, rejecting business class champagne. Trading my champagne for another soft drink, I return to the cockpit and wait for more communication from Sydney.

We've been on the ground for about ninety minutes. The passengers are secure in the small airport terminal, and Qantas is working towards a rescue plan. I'm still not getting a mobile phone signal.

My role as the captain hasn't been relieved; I'll probably be required to continue to exercise my authority in the hours ahead. A quote from General George Patton rises out of my memory: 'Do more than is required of you.'

I've done a lot already. Chunks are being torn off me as I absorb the strain and confronting scenes – but there's much more for me to do. I'm still in command and this drives me to stay focused, to keep thinking, to be compassionate and to ensure our rescue is successful. My military experience has prepared me to accept this responsibility in the face of fatigue, frustration and extremis. I'll continue to lead.

# 16.

My career in commercial aviation has been anything but boring and routine. I've been challenged by many equipment failures and severe weather events, and I've always managed them appropriately and efficiently. Many have been classified as extreme events: an uncontained engine failure with severe vibration; airframe and engine damage caused by severe hail on take-off; severe icing at altitude resulting in engine damage; and numerous fume events. Plus the myriad non-routine events that all commercial pilots must deal with in the course of their duties.

I've learned from these events, but none have generated the body response or trauma that this one has. This scenario involving computers, denial of control and potential mass casualties is at a different level. It seems we've survived a science-fiction scenario, a No Man's Land of automation failure on an unprecedented scale. I wasn't in a position to guess what was happening or to make up a procedure. But I had to make some big decisions – extreme decisions to cater for an extreme event. Looks like my shit-magnet reputation will soar to new heights after this.

I know my crew and I will be thoroughly questioned in corporate headquarters. I must try to make sense of what I saw, how I reacted and what I said. My compliance with company and Airbus procedures will be heavily scrutinised by safety and fleet management pilots.

Have I missed something?

I mentally review the sequence of failures and bad air data presented to me in the form of unreliable speed and altitude on my primary flight display tapes. But I can't access the data that was generated from the ADIRU and presented to the aircraft's computer systems. Perhaps the computers activated some 'protections' that are built into their software – but surely we should have been warned that these protections would be activated. Why hasn't the post-flight report or the status summary indicated which modes were activated and for what reason? Like a bad partner, the computers systems went crazy then stopped communicating with me.

This complex failure scenario was not raised during the Airbus conversion course. I doubt that even the Airbus test pilots have been exposed to this scenario.

One thing is certain: the computers blocked my control inputs. For a pilot, loss-of-control is the ultimate threat. It's our job to control the aircraft, and if computers and their software, by design, can remove that functionality from the pilot, then nothing good is going to come out of that.

I know our decision to divert to Learmonth, instead of continuing to Perth, will be scrutinised. But what risks would we have taken if we'd flown the extra distance to Perth? I shudder at the thought.

I find hollow solace from the doctor in *Two and a Half Men*: 'If you put a tuxedo on a goat, it's still a goat.'

*

The time is approximately 1630.

The satphone is ringing again – it's someone from the IOC in Sydney. The conversation concerns the rescue plan. Two aircraft will be sent to Learmonth, a 767 and a 717, to airlift the passengers and crew who are able to travel. The first aircraft will arrive around 1715, followed closely by the second. The Royal Flying Doctor Service has allocated five aircraft to assist in airlifting the more seriously injured from Learmonth to Perth.

I brief my crew on the rescue plan while encouraging them to eat and drink to help them through this post-trauma phase. They're still struggling with the shock.

Now it's time for me to share the rescue information with my passengers in the terminal waiting area. I try to compose my briefing to the 290-plus passengers on the walk to the terminal. Inside, I'm given access to the PA microphone but the staff member isn't sure how to use the system. Finally a local police officer offers me the use of his megaphone.

*Okay, this is going to look interesting to the passengers*, I'm thinking, but I accept his offer.

Part of the terminal is being used as a staging area for the more seriously injured, while the other half is loaded to the brim with everyone else. Armed with my megaphone, I ask everyone to come closer and suppress a rush of emotion as my passengers form a semicircle around me to hear my news.

'Ladies and gentlemen, can I have your attention? I have some information that I want to share with you regarding our rescue from Learmonth. I have been informed that two aircraft will be arriving soon to take us all to Perth. I know you all have belongings on the aircraft and I'm sure they will be returned to you soon. The hardest thing for all of us to do now is to be patient, and I ask all of you for your patience now.

'As far as what happened during our flight, all I can say is that I don't know what happened but I tried to stop it. But we are here now and we are safe. I know Learmonth is an isolated place, but the runway looked pretty damn good to me as we rolled out on final approach for landing! So please, be patient and we will all be out of here soon. Thank you.'

As I mention seeing the clear runway on final approach, the entire crowd erupts in cheers and applause. Someone to my right is filming my speech and capturing the ecstatic response of the passengers.

We've survived a profound near-death experience together, and now I can't help feeling we might as well be marooned on a desert island; we are truly in a survival situation, and I'm still the man responsible for their safety.

Everyone is frightened, concerned and traumatised. The loss of mobile phone coverage prevents them from contacting loved ones waiting for them in Perth. I spare a thought for my daughter, knowing she'll see this event on the news when she arrives at her next hotel in southern Europe; I hope she gets my text message.

I spot Fuzzy cocooned in a stretcher on the floor. Straps secure him and he has an enormous brace on one leg. My heart aches, and my body is burning at the sight of Fuzzy and his injuries. I kneel down and lock hands with him. Help is on its way, I tell him. He'll have to be airlifted out by the Flying Doctors to Perth for treatment, without the support of his crew.

Diana Casey and her two daughters are awaiting transport to the Exmouth Medical Centre. The girls aren't injured – they kept their seatbelts on – but they are crying quietly. Peter has been ambulanced to the centre already and Di is waiting her turn.

I'm having a difficult time absorbing all the pain, suffering and uncertainty enveloping my passengers and crew. I must carry this burden, but I'm bleeding emotional energy to maintain my strong face of command. I'm still in the knife fight, and the slashes are hitting home with every new challenge I face.

The local airport staff are organising food and drink for the passengers and crew, and I marvel at their presence and compassion. The sleepy town of Exmouth has mobilised; many have volunteered to assist us at the airport.

When I return the megaphone to the helpful police officer he tells me that the aircraft will soon be under police guard, so the crew will have to vacate it; I should go over to let them know. While we're talking, a charter aircraft arrives on the apron, and WA police officers assemble on the tarmac. The first air ambulance has also landed, and its propeller is winding down while an ambulance from Exmouth Medical Centre approaches to transfer injured passengers. Help is nearly here.

*

The sun is getting low on the horizon as I climb the stairs to the aircraft. Some of the crew are still there, and we watch silently as the skies turn purple and red with the setting sun. I take only one photograph on this day: the sunset with the aircraft in the foreground, a serene scene as the mauve-hued clouds reflect off the white skin of the fuselage. The beauty of the sunset masks the carnage and violence that took place inside. There's a collective, sombre pensiveness among the survivors with the reality that the day's trial may have prevented us from ever seeing a sunset again.

Police officers approach, and a sergeant cordially introduces himself to me. He's tall and dark-haired with a neatly trimmed moustache, and tanned from the tropical sun that bakes his base in the small town of Karratha. He explains that the aircraft is now in lockdown and asks the crew to leave.

I jokingly ask if the aircraft is a crime scene

'Something like that,' he replies.

Reluctantly my crew gather their belongings.

I'm permitted to stay onboard to monitor the satphone, but I accompany my crew to the tarmac in search of a suitable private area for them. One of the airport staff offers some baggage carts near the tarmac for the crew to use. I want them to stay together and away from the main terminal. The effects of the post-adrenalin response are taking their toll. All stare blankly as they sit on the carts; one is shaking quietly but not because she's cold. They need some privacy now and to support each other until the airlift is organised.

Across the tarmac, a van arrives, loaded with sandwiches and drinks. I grab a ham-and-cheese sandwich and a bottle of water. As I steal a quiet moment to eat, I spare a thought for Ross and the three remaining cabin crew members from the Auckland base. They're young and this experience has brutally stolen their innocence. But possibly this innocence has been ripped out of everyone onboard, old and young alike. Sure, shit happens in aviation, but this disaster is shit of a different magnitude. If an aircraft accident involved this number of serious injuries, you'd expect to see the hull broken up, twisted and burning on the ground – not sitting serenely on a tarmac, intact, with a picturesque sunset in the background.

On the tarmac, another air ambulance arrives and shuts down its engine. I'm still unaware of the numbers of seriously injured passengers and the nature of their injuries, but I know the rescue operation is in full swing. Twenty passengers are declared seriously injured and are to be sent to Exmouth Medical Centre to be examined for spinal injuries and ensure fitness for the flight to Perth via air ambulance. On-scene reports state that all are being categorised as having spinal injuries, including one of the pilots, and there are two cases of broken femurs.

I make my way back to the aircraft and join the senior sergeant beside the left entry door. We admire the colourful light show of the setting sun, and I can't help but think the scene should feature an esky full of cold beer at our feet.

Matter-of-factly, he lets me know that a police photographic team are onboard recording images of the damage. I excuse myself and find them in the rear of the aircraft, moving methodically through the cabin. They're capturing images of the punctured holes above the passenger's seats, the busted-up galleys, the missing ceiling panels and luggage lockers, and the bloodied bandages and pillows strewn throughout the cabin.

One of the toilet doors is open, and they take photos of the uprooted toilet that lies on its side. There's blood on the walls here, too. I didn't see this before. I suppress another shudder at the thought of the unfortunate occupant.

The unmistakable roar of jet engines in reverse thrust signals the arrival of the first rescue aircraft. It's 6.18 p.m. The 767 backtracks along the runway to be parked at a remote area. The tarmac is full to capacity with our A330 and various smaller aircraft, so the 767 must park somewhere else on the airfield. I've been told this aircraft contains customs and immigration officers, as well as the Qantas assistance team.

*Maybe I can relax a little now*, I think to myself.

But no, there's still too much more to organise and oversee. I channel my resolve to 'stay up' … We are not out of here yet.

As if on cue, a plain-clothes detective approaches me and cordially introduces himself. He's young and blond, with an innocent but official air to his posture. He says he needs me to make a statement.

I'm back on high alert. 'What will this be used for? Is this a criminal investigation?' I understand he's only doing his job, but it's highly irregular for an airline captain to be giving an official statement to law enforcement after an emergency landing.

A terrible thought flashes through my mind. Has a passenger lost their life due to their injuries? My face reddens and my heart rate accelerates again.

I consent to the interview process, and the detective takes out his yellow legal pad. At first I answer some basic background questions such as age, employment and flying experience, but when we get down to the nitty-gritty of the nature of the emergency, it becomes more complicated. I explain that, for an unknown reason, the aircraft pitched down.

'So, the plane dived, is that right?'

'No, it "pitched" down. There was altitude lost in the recovery, but this isn't a dive.' I pivot my hand to mimic the manoeuvre.

'Okay. What caused the aeroplane to do that?'

'Unknown.' That might sound a bit fishy, but I'm still in the dark. I can't truthfully state what caused the aircraft to malfunction; I can only tell him what I did and why I did it. I direct my explanation to the computers and automation failing, but this is too sci-fi for the detective.

After about thirty minutes of our toing and froing, he records a statement that concludes with something like this: 'I declare that this statement is true and correct to the best of my knowledge and belief and I have made this statement knowing that if it's tendered as evidence, I will be guilty of a crime if I have wilfully included in this statement that is false or not true.'

Once the detective is satisfied, I retreat to the sanctity of the flight deck, feeling more like an accused shoplifter at the local 7-Eleven than a captain who has successfully delivered his passengers and crew to safety. At least I haven't been charged or hand-cuffed.

*

The sun has set and the airfield is cloaked in darkness. A quick glance at my watch shows the time is 1850.

As Flying Doctor air ambulances prepare for departure, a minivan arrives at the terminal. Two of its occupants cross the tarmac, heading for me.

A police officer standing guard at the entrance speaks briefly with my visitors before granting them access. They murmur a few 'holy shits' before they enter the flight deck. They're company ground engineers from the Perth base; one is an avionics specialist I recognise from previous flights through Perth.

I go through the sequence of events with them, highlighting the many failures and faults, the ineffectiveness of my efforts to stop the aircraft's manoeuvres and the scrolling faults. I may as well be describing a brutal mass murder, judging by their widening eyes and gaping mouths. I explain that the destruction in the cabin was caused by the bodies of the passengers and crew being catapulted into them and that, perhaps, I'm pretty lucky to be giving them a face-to-face debrief.

They look at me as if they've just met Lazarus. This explanation is unknown territory for them too; there is no history of this happening before on an Airbus.

The avionics engineer starts to access the maintenance computer and the post-flight report through the computer keyboard. He asks me questions as he sees the array of faults and failures summarised on the screen. 'Did you action this fault?' 'No, it didn't latch.' 'How about this one?' 'No, none of those latched.' I summarise the faults that did latch, which allowed us to reconfigure, and finish by expressing my view that the aircraft systems experienced a complete crash and we were understandably confused.

As the engineers access the electronics equipment area under the floor, I hear the sound of jet engines roaring in reverse thrust as another rescue aircraft touches down.

# 17.

Somewhere in Europe, my daughter and her boyfriend arrive at their next destination, check in to their hotel and settle into their room. My text message has my daughter on edge as they turn on the television to the English channel for CNN. Plastered on the screen is the 'breaking story' of the Qantas flight from Singapore to Perth, and the ensuing emergency landing in remote Western Australia with multiple casualties.

'Dad was flying to Perth today. It must be him!'

Her boyfriend offers his feeling that someone else is involved in this accident, not me, but she has received a text message like this before from me and is certain I'm involved in this disaster.

There isn't much information due to the remoteness of Learmonth. She tries to call my mobile phone but it diverts to message bank. Her concern grows as fast as her sorrow; at least she knows I'm 'okay' from my text message. She's desperate for more information but it will be hours before I can speak with her.

*

At the Exmouth Medical Centre, the staff are working quickly and methodically. The facility is working at overcapacity, with six beds to cater for the twenty seriously injured. As they arrive, they're X-rayed to see if there's evidence of spinal injury.

The Casey family is here, finally reunited and resting after their examination. Diana and Peter have been cleared of spinal injury, but they'll require more tests to determine the full extent of their injuries. Diana's shoulder is now quite stiff, and Peter's face is swelling and changing colour in response to his violent collision with the ceiling. They're resting in adjoining beds, their daughters nearby, as they await information on their airlift arrangements to Perth, courtesy of the Royal Flying Doctors. There's calm in the eye of the storm.

Then a nurse tells them that their daughters must return alone to the airport for their airlift flight. The air ambulance doesn't have enough room to transport them all together. Diana pleads with the medical staff to make other arrangements to keep the family together, but her request can't be granted. Her distraught daughters are in tears as they're led away. One of the assisting female doctors, who has been using her own private vehicle to ferry passengers to and from the medical centre, volunteers to take Diana's girls back to the airport.

Soon Diana and Peter are loaded into an ambulance for the thirty-minute drive to Learmonth Airport. Although the medical staff have cleared them to fly, Diana can't move her shoulder, despite some pain-relief injections, and Peter's face has swelled to the point of closing his left eye. He's had a tough twelve months battling a serious medical condition, and Diana refuses to leave his side.

*

The Qantas Link 717 arrives. It shuts down its engines and disgorges its load of support personnel.

Another wave of activity approaches. Two representatives of the Australian Transport Safety Bureau (ATSB) have arrived. It turns out the aircraft has been placed in lockdown by the ATSB to facilitate their investigative processes.

Like the engineers, the investigators are visibly awed and horrified. The destruction in the cabin looks like a goldrush-era saloon after a bar-room brawl, unlike anything they've experienced before in their careers as safety investigators. I introduce myself to them as they solemnly commence their investigation. I don't offer any statement at this time.

I gather my bag and make another quick scan around the flight deck for my uniform tie, missing since the first pitch-down. It's nowhere to be seen; it has been swallowed into the bowels of QPA.

I hesitate near the forward exit door, capturing the carnage in my memory. It's fitting that I'm the last to depart QF72. *Fortunately, I didn't go down with the ship*, I muse, *but too damn close for comfort.* As I walk down the steps, I flash back to my navy days. Whenever a senior ranking officer departed the ship via the gangway, there would be an announcement and a ringing of the forecastle's bell. I'm leaving my ship now, my command. *Ding-ding. Ding-ding.* Captain Kev, departing.

The tarmac is full, squeezed to capacity with three more air ambulances, a Westwind charter jet that has flown the customs and immigration staff in and the 717 rescue aircraft. The aircraft are parked, primed and ready for action.

Near the tarmac I join Pete, who's talking to the captain of the 717. He's astonished as he listens to our description of the Airbus A330's systems meltdown. 'You know what they say … if it ain't Boeing, I ain't going,' he quips.

I'm surprised to see two of our passengers occupying the tarmac. Sitting in their cages are our live-cargo dogs, somehow extracted from the bulk cargo hold despite a lack of appropriate equipment; the Qantas engineers have improvised. 'Who let the dawgs out?' I sing as Pete rolls his eyes. I'm relieved they're in the fresh air, tails wagging and patiently waiting for their ride to Perth.

There's a lot of activity on the tarmac but not much is happening. The 717 captain's forehead is furrowed in concern as he checks his watch. 'I was supposed to be airborne at eight o'clock. I told the assistance team that if I'm not loaded and airborne soon, I'll run out of duty time.' If he can't fly because his duty period is exceeded, then accommodation for 150 people will have to be found in Exmouth – a tall order.

Pete and I take action. We emphasise to the company assistance team the urgency of this situation. Minutes later, we're relieved to see passengers making their way towards the waiting 717. I stand in front of the aircraft and make eye contact with the young captain in his cockpit. I give him a thumbs-up; he smiles and returns the gesture.

*

In Sydney, Captain John Killingback is informed of the QF72 mayday. When he's told that I'm the captain, he springs into action. We are neighbours and close friends.

He first calls the president of our pilots' association and offers to travel to Perth immediately to act in a supportive role for the pilots. The president agrees and tells him to get to the airport as fast as possible to travel on the last flight to Perth – it departs within the hour!

As John packs his bag with essential items, he calls the chief pilot to offer his services as a direct representative of flight operations. The chief pilot also agrees and confirms he'll assist in John's travel.

After giving his wife a quick kiss, John is out the door and racing to the airport, thirty minutes away by car. He arrives at valet parking in a cloud of smoke, throws his keys to the attendant with 'I'll be back in a couple of days', and sprints towards the departure gate, bounding up the escalator. He doesn't look back at the perplexed valet staring down at his keys.

Alerted by Qantas operations, the ground staff are waiting for John at the departure gate. They usher him down the aerobridge and through the entry door. He's the last to board. As he nonchalantly waves to the pilots in the cockpit, the door is secured behind him and the plane commences its departure.

*

Another friend and colleague, Captain Phil Paterson, is now back in Perth after completing his day of flying. He heard our mayday call while he was airborne. He, too, has communicated with the chief pilot and flight operations management. Like John, he offers his participation in a supportive role for the pilots once we arrive in Perth.

The deputy chief pilot instructs Phil to do whatever it takes to help the pilots.

*

The time is 2133.

It's getting late. We watch as the 717 departs for Perth. At the insistence of the special assistance team, a few of my crew are on board, although we are all reluctant to be separated. The plane is dispatched with only minutes of pilot duty time to spare.

A minibus is shuttling passengers to the 767, parked at the far end of the runway. I feel a bit like a shepherd tending my flock, ensuring their journeys are completed.

There are three more air ambulances on the tarmac now, waiting for their fragile cargo. I admire the operation of the Flying Doctor flights and their crews, and I'm in awe of the efforts of the many medical staff mobilised to support our rescue. Someone, somewhere, has their finger on the pulse to ensure this side of the rescue is progressing well.

The terminal is empty with the exception of the WA Police officers and local airport staff. The tarmac is still buzzing, and

QPA holds the dominant position, parked on the right-hand side. From my angle, it looks like a venomous insect with its highly curved wing and pointed tail, ready to inflict more damage on an unsuspecting prey. We were its prey.

This is the point that I walk away from it.

My crew is solemn and silent. We're leaving this nightmare behind. Maybe I can relax a little … maybe.

As I climb the stairs to the 767 behind my crew, I glance one last time at Learmonth Airport. I'm the last to board our rescue aeroplane.

When I enter the business-class area, I'm met with irritated stares from the passengers. After surviving the horror of the pitch-downs, they've been here for hours, waiting to depart. The patience I asked for when we landed has eroded; it's been a long and emotional day. I'm grateful when these passengers offer sporadic claps and cheers as my crew and I move towards empty seats in the rear of the cabin.

Through the vibration of the cabin floor, I feel the forward entry door close with a *thunk*. I tighten my seatbelt and look out the window as the second engine is started. One of the PC-12 air ambulances is on the runway, wing lights illuminated and white strobe lights flashing, ready for take-off. Who occupies the stretchers in the back of that aircraft, I wonder?

The 767 is taxiing forward and onto the runway. The strobe lights are beating quickly, illuminating the area where I'm seated. I force myself to stay awake to watch the tarmac disappear beneath me as we depart for Perth. The pilots turn onto the runway centreline, the 767 rotates, and we are airborne.

As Learmonth disappears behind us, a massive weight pushes me back against my seat. It isn't from the force of the take-off – I'm overwhelmed and emotionally exhausted. My body is sinking into the seat, and I won't fight it now. I can't keep my eyes open any longer.

Captain Kev, departing.

# 18.

'Hi, honey. It's me.'

'Dad! What's happened?! Is that your flight I see on CNN? I've been trying to call you all day!'

She sounds quite angry with me, and I can understand her frustration and fears.

'Yes, honey, it was me. I sent you a message to say I was okay … Did you get that?'

'Yeah, but why didn't you call me?!' Here voice is choked with tears of relief mixed with frustration.

I explain the lack of mobile phone signal at Learmonth and that I've been a 'bit busy' until now. I assure her I'm safe in Perth and wasn't physically injured.

I promise to call her later when I get to the hotel. I ask her to think about returning early from her trip, as I believe Qantas will offer to assist her in coming home. I tell her I love her and ask her to try to settle down, now the worst is over.

It's been less than an hour since we touched down in Perth at one in the morning. The passengers clapped and cheered as the aircraft decelerated. I twisted around and made eye contact with Pete and Ross; they responded with wry, weary grins and exaggerated exhales. As the passengers filed out of the plane, some looked back and waved at the crew in thanks. I wondered how soon these travellers would venture onto an aircraft again after this ordeal, if ever. I doubted that Airbus would be at the top of their preferred aircraft list after this frightening day.

I lingered in the back with my crew until all the passengers had deplaned. On my way out, I thanked the cabin crew of this rescue flight. Reaching the main cabin door, I looked back to ensure all my crew had disembarked before I walked out into the bridge.

I was the last QF72 survivor to leave the 767. My duty as the captain of my flight was then complete.

My friend Captain John Killingback, who had rushed here to help me out, was among the throng of airport staff lining the aerobridge. I smiled and walked straight over to him, relaxing a little. He's very good at handling people and isn't afraid of stepping on toes to achieve an outcome.

We were led to where two trauma specialists and a GP were waiting to assess us.

The GP started with me. 'Are you physically injured?' he asked.

I told him I was fine, but that some of my crew may need medical evaluations at a hospital.

*

When the minibus delivers my cabin crew to their hotel, I'm relieved to see the media aren't in attendance. I hug each crew member and promise to see them later in the day.

Soon enough, the minibus is pulling up outside my hotel. The media aren't here either. All I can think about now is a shower, but we have no clean clothes – our bags are still in Singapore. John has invited me to his room to talk and have a beer, but my clothes are quite ripe and I have to freshen up.

I stay under the shower a lot longer than usual. I finish with a blast of cold water to cool my flushed and burning body. The difficult part now is, what do I wear?

The phone rings; it's John asking if I'm coming over. 'Yes, yes,' I say, 'calm down.' I take a whiff of my uniform shirt and decide it should be buried or burned. My underwear isn't in any shape for recycling, either. I find a terrycloth bathrobe in a closet, but it's very short, more suitable for lounging around at the Playboy Mansion than having a beer with my fellow pilots. Oh well, it will have to do.

I feign a shy, sheepish manner as I saunter into John's room and try to sit in a position that will prevent my block-and-tackle from escaping my risqué robe. John, in his most parental voice, rebukes me with an 'Oh geezus, Kevin!' and we all have a laugh at my expense. Someone hands me a beer. It's ice-cold but tastes horrible – once again, alcohol doesn't sit well with my palate, as with the champagne in Learmonth. I put it down.

'What? You're not going to drink your beer?!' Ross exclaims in mock horror.

This is sacrilege in Australia and my colleagues take great delight in chiding me about it, along with my comical robe.

We settle down and try to re-create our flight briefly for John's benefit. I know the Sydney managers will ask him for his comments later in the day. He listens carefully as we share our most critical recollections of the flight, but it's four in the morning and I've been up since six the previous day … seems like a lifetime ago.

I give John a manly hug in thanks.

He doesn't skip a beat. 'You're not my type.' Especially in my fetching terrycloth robe.

I know he'll be traumatised by the memory of me in that robe, and I'm eternally indebted to him for his assistance on the worst day of my life. He's a good mate in every way.

I collapse into bed. The pillow has the gravitational pull of a black hole as I glance at the time: 0430.

*

Diana and Peter Casey are finally airlifted to Sir Charles Gairdner Hospital in Perth, where Peter, along with Fuzzy, is admitted to intensive care.

Somewhere in the city in a hotel are Peter and Diana's daughters. As soon as she's discharged, Diana goes in search of them, walking the eight blocks to their hotel. A quick call to the room has them down in the lobby in an instant for a tearful reunion.

*

I don't remember dreaming, but something in my subconscious wakes me like an electric shock. I bolt upright, quickly scanning my room to get my bearings. Oh yeah, I'm in Perth, safe in the hotel. It's 6 a.m. The sun is up and my blackout curtains aren't doing their job very well. I can't go back to sleep; my body is still hot and I feel like I'm in a vice.

I try to make sense of the past twenty-four hours. I'm feeling exhausted but, like any professional pilot, I commence a self-critique of my actions yesterday. Could I have done anything different, anything better? Have I completed the flight in accordance with the aircraft operating manuals and company policies? I answer 'yes' to all these questions.

I also have a deep, burning pride for the heroic actions of my crew in the face of this cataclysmic accident. There's no company training that covers this type of scenario, and somehow they adapted and performed admirably.

I know all onboard will be adversely affected with the confronting reality that they were faced with a life-threatening event. Despite my history of flying aeroplanes from the decks of aircraft carriers, I must swallow the bitter pill of reality and conclude that this is already affecting me too. Even though we've survived something truly life-threatening, I can't extract any joy from this reality.

When I vividly recall the image of the ocean in my windscreen and my helplessness as I tried to arrest the computer's manoeuvres, my threat response fires up anew. My memory replays the images of the injuries to my passengers and crew, the bits of scalp dangling from the broken plastic, and the children crying in their parents' arms with bruised and bandaged heads. That recollection activates something in me, and I start to sob. I sob like I never have before. Is it the post-adrenalin recovery sequence? Or is it a jab of self-pity?

I've dedicated my entire life to being a pilot. Now, with regret, I see my life's work sailing off into the distance. I'm not sure if I can swim fast enough to get back onboard.

This experience will change my life and my career; I knew it after the first pitch-down. The body response I experienced during the dives, and after, was enormous. I feel an overwhelming sense of loss. I don't fight it; I'm alone in my room and I allow my body to rack me with this convulsive sobbing until it subsides. I accept this as part of my recovery, but I'm sad to have been thrown over a threshold into unknown territory; I've been catapulted into an emotional No Man's Land.

Throughout my flying career, and especially while flying in the navy, I worked hard to master the art of compartmentalisation, but I'm concerned this event may be too large to jam into any of my existing compartments. In that hotel room, my memory is like a videotape on constant replay, and I'm finding it hard to stop, to pause or suppress its content.

I force myself to take a few deep breaths and try to settle down, but for the first time in my adult life I feel derailed and confronted by something that may be outside of my experience to cope with. I retreat to the shower again, to cool down my body, to try to wash away my anger and camouflage my tears. This is not me, but maybe it's the new me.

# 19.

Early the next day, Captain Phil Paterson calls me. He's in Perth to assume a support role for us, and he suggests breakfast. Great idea, but I have nothing to wear. I borrow a t-shirt from John, stencilled with 'In Dog Years I'd Be Dead', it matches my blue uniform trousers but it's a bit too small.

'Looking quite fetching in your tight t-shirt and matching uniform trousers, Kev,' Phil dryly observes.

'Yeah, well, John needs to go to the gym more and buy some bigger shirts for moments like these. You should also know that I am "commando" … my underwear is in no state for use and these trousers are a bit itchy without a layer protecting my bits.'

My two colleagues razz me about my outfit all the way to the café.

The first order of business is to update Phil on the events of the previous day, including the difficulties at Learmonth and getting the crew proper medical care here in Perth.

We then learn about our schedule for the next twenty-four hours.

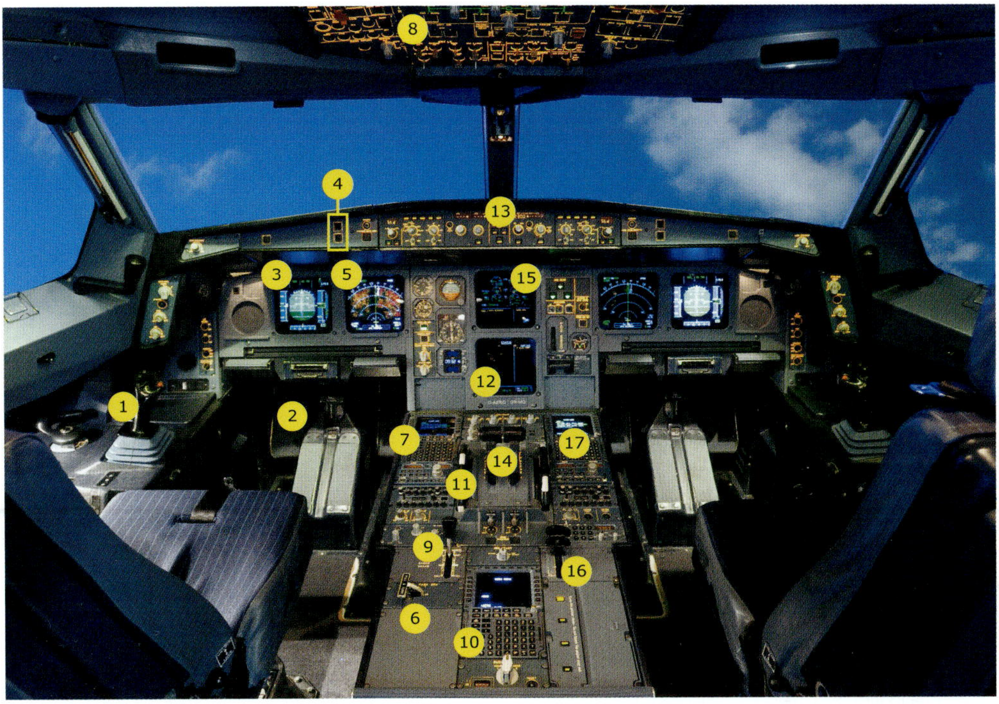

The 'glass cockpit' of an Airbus A330. *(Norbert Michalke/Alamy)*

1 control sidestick
2 rudder and brake pedals
3 primary flight display
4 master caution and master warning lights
5 navigation display
6 park brake
7 flight management computer 1
8 aircraft systems control panels
9 speed brake
10 flight management computer 3
11 trim wheel
12 aircraft systems display
13 speed, heading and altutude controls including autopilot
14 engine thrust levers
15 engine/warning display
16 flaps lever
17 flight management computer 2

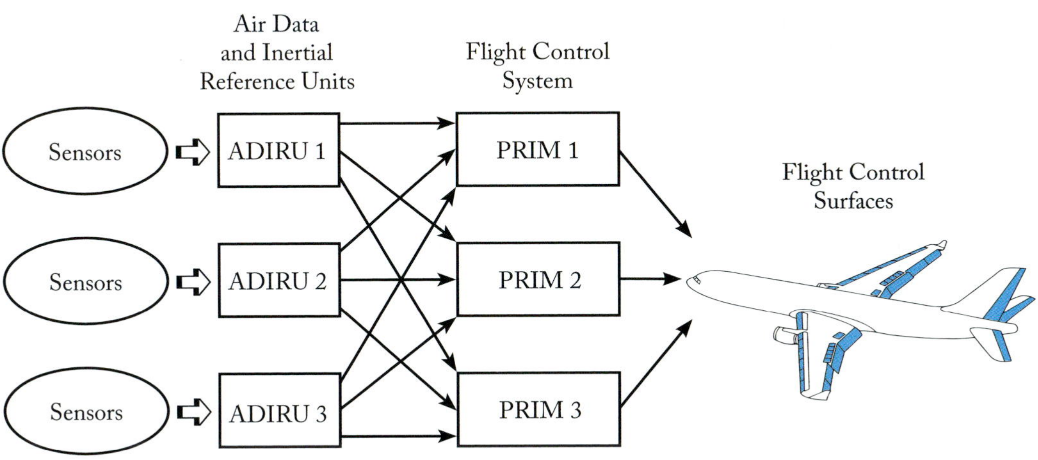


The 'transparent' interface between the aircraft's external sensors, ADIRUs and the flight control computers. There's a lot going on between those units and the lines connecting them.

September 1977, Training Squadron 6, NAS Whiting Field, Florida. A young Ensign Sullivan starts his path to earning his wings. A new flight suit, my first helmet with my lucky shamrocks and full of motivation. *(author collection)*

Wings Day, September 1978. A proud mother has pinned Navy Wings of Gold on her son's chest. NAS Chase Field, Beeville, Texas. *(author collection)*

Coming aboard the 'Lady Lex' (CVT-16) as viewed from the back seat of a T-2C Buckeye.
*(author collection)*

Manning up for a morning launch on board USS *America* (CV 66). On work-ups prior to first deployment with fighter squadron VF-114, January 1981. *(author collection)*

Above: Returning from a dawn patrol in the northern Sea of Japan. *(author collection)*

Left: Transit of the Suez Canal, October 1981. Egyptian president Anwar Sadat had been assassinated days prior and the USS *America* was on full alert for a potential terrorist attack. This was the *America*'s second transit of the Suez. *(US Navy)*

Fighter Squadron VF-114 'Fighting Aardvarks' on board USS *Enterprise* (CVN 65), 1982. Our orange flight suits match our squadron mascot 'Zot'. *(US Navy)*

My radar intercept officer, 'Breeze', took this shot of a Tupolev TU-95 'Bear' in the vicinity of USS *Enterprise*. Escorting nosey Soviet aircraft was a routine occurrence for a Tomcat driver during the Cold War. *(US Navy)*

Above: On exchange with the RAAF's No. 2 Operational Conversion Unit. We were in transit to Darwin and have dropped into Alice Springs for fuel. My wingman and I pose with 'pistolas' drawn while we refuel. *(author collection)*

Left: My final flight in a fast jet. I reflect on my career and my lost comrades, reluctant to leave the cockpit of my Mirage 3. Speed Is Life and I will miss it. RAAF Base Williamtown, New South Wales. *(author collection)*

First Officer Sullivan (fifth from left) with the Qantas technical team, picking up a new 747-400 (VH-OJJ) from the Boeing factory on a wet day, February 1991. *(author collection)*

Operating a 767 as a first officer, heading north to an exotic Asian destination. We still need to use our charts to check our navigation system. *(author collection)*

VH-QPA in quieter times. *(CC-BY-SA-2.0 Masakatsu Ukon)*

QF72 on the parking area at Learmonth Airfield after the pitch downs and emergency landing. We were lucky the external stairs were available for our disembarkation. *(Kristin Anderson/AAP)*

My operational debrief to the crew on the ground at Learmonth; trying to explain the unexplainable. They were all so heroic. *(F/A Rory)*

The West Australian police have asked the crew to vacate the aircraft. CSM Lisa asks for clarification while the crew look on. *(F/A Rory)*

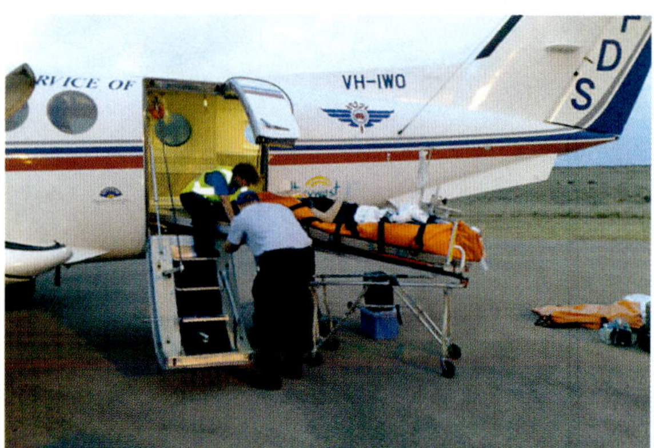


The first of the severely injured are flown out of Learmonth to Perth by the Royal Flying Doctor Service. The medical response of the RFDS and medical teams at Exmouth was phenomenal. *(F/A Rory)*

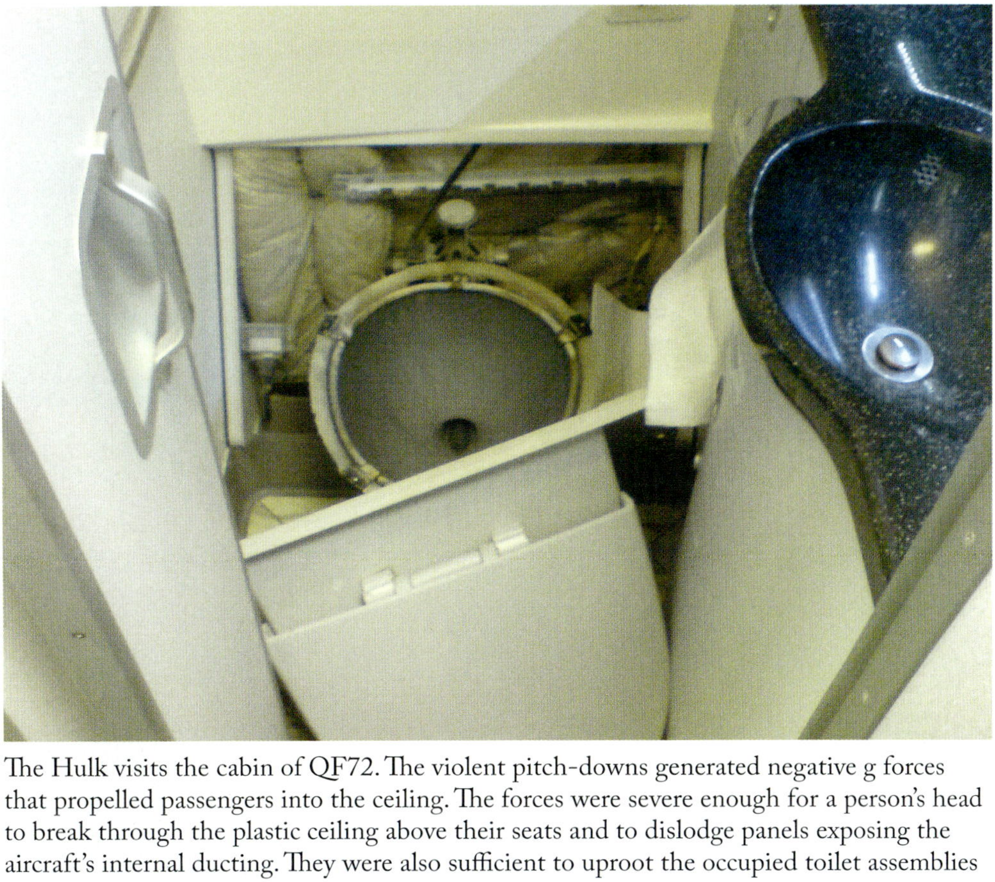

The Hulk visits the cabin of QF72. The violent pitch-downs generated negative g forces that propelled passengers into the ceiling. The forces were severe enough for a person's head to break through the plastic ceiling above their seats and to dislodge panels exposing the aircraft's internal ducting. They were also sufficient to uproot the occupied toilet assemblies from their mounts. *(F/A Tasha)*

More broken panels, causing more broken bodies. *(F/A Tasha)*

The rear galley where the negative g forces were highest. Note the deformed stainless steel bar on the left side of the counter. This was caused by flight attendant Fuzzy Maiava's body as he was slammed off the ceiling during the computer's first pitch-down manoeuvre. *(F/A Tasha)*

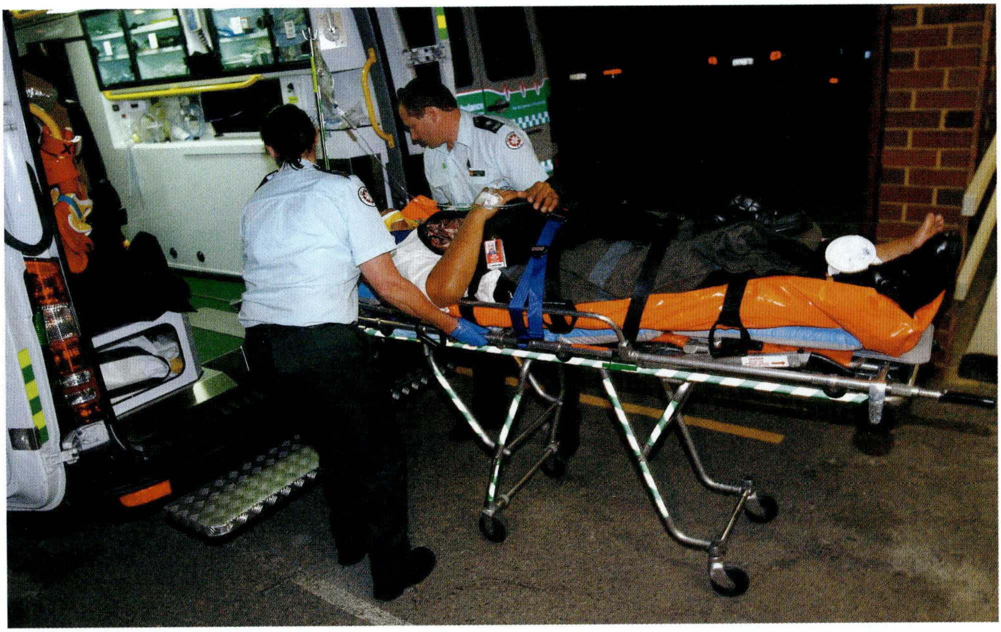

Fuzzy is loaded into an ambulance in Perth after being airlifted from Learmonth by the RFDS. *(Paul Kane/Getty Images)*

A tired but uninjured QF72 passenger decsribes the pitch-downs to awaiting media at Perth Airport. *(Paul Kane/Getty Images)*

Fuzzy Maiava struggles to repair his broken body and broken life. His right knee bears the scars of his knee reconstruction. He has difficulty managing chronic pain, PTSD and extensive soft tissue damage caused by his injuries. *(Chris Skelton/Stuff)*

Air Transport Safety Bureau director of Air Safety Investigation Julian Walsh holds the flight recorder from VH-QPA during a press conference in Canberra. *(Alan Porritt/AAP)*

My last day at Qantas Airways. I lost my battle with PTSD and was medically terminated. Captain Matt Hicks, the hero pilot of QF32 (when an uncontained engine failure crippled an A380 flying out of Singapore), congratulates me on completing my commercial flying career. *(author collection)*

A pensive portrait from Fairfax Media in support of their article entitled 'The Untold Story of QF72: What happens when "psycho" automation leaves pilots powerless'. *(Louie Douvis/SMH)*

We'll depart Perth tomorrow for Sydney, but first we must be medically cleared to travel as passengers. We're to attend Sir Charles Gairdner Hospital for a medical examination. Secondly, our luggage from Singapore will be flown down on the inbound Qantas flight and delivered to the hotel. Third, permission is given for the three of us to purchase essential clothing and toiletries at company expense.

On the way, we stop to buy some appropriate clothing. Thankfully, at the hospital the staff are expecting us, and we're led into a triage area. I'm questioned by a doctor about physical injuries and confirm I don't have any.

'How are you feeling, Captain?'

'Like I've been run over by a bus.'

'Are you physically injured?'

'No, but I had a huge body response after the aircraft injured everyone on my flight, and this morning too. I got a good look at the Indian Ocean yesterday and that wasn't very pleasant.'

He nods gravely in understanding and mentions a term I will hear for years to come: post-traumatic stress disorder (PTSD). He's just making an assessment for fitness to travel, not to offer treatment, so he concludes my examination and annotates the 'injury' section of his medical release with 'PTSD'.

*PTSD … What the hell?*

I don't know much about PTSD. I knew it was called by other names like shell-shock, battle fatigue and combat fatigue, but now my event has invited me into this club. I am aware of adrenalin response through my military experience but this is very different.

In the light of the doctor's assessment, I see that a lot of things are connected: extreme body response, near-death experience, the trauma of witnessing injuries, good champagne tasting like copper, and fight-or-flight chemicals. A four-letter acronym that will have so much more associated with it in the years to come.

We depart the hospital with our clearance to travel and return to the hotel. I learn that some of my crew are departing Perth later in the day for their homes; some in Brisbane and Melbourne, one in Singapore, and my three cabin crew from Auckland. Those three were worried they'd have to operate as crew members directly after this event but, fortunately, they're returning home as passengers. I head down to their hotel and try to see them before their departures.

This bomb-burst of the crew departing for home seems wrong but it's typical, in some ways, of the airline business. This is the way most trips finish because everyone is keen to return home. This isn't a typical trip, though. I show a strong face, but privately I'm sad and concerned as we exchange our goodbyes.

Adam Lloyd, the Airbus assistant fleet manager in Sydney, calls. He isn't a pilot, and his role is to manage the abundance of administrative issues that confront a pilot in the course of their employment. I've known him for many years; he's truly a compassionate man. Now he's offering assistance.

We discuss, again, the events of the previous day and he listens closely to my recollection. As a matter of course, my daughter is listed as my emergency contact, and Adam shares his unsuccessful attempts to contact her and provide information. I inform him that she's overseas on holiday, returning in a week or so.

'Does she want to come back early?' he asks. He knows how close we are; I used to take her on trips with me when she was younger.

I know in my heart that, right now, I want her to be physically closer to me than Europe.

'Yes, please help her to come home,' I tell Adam.

Qantas flights from Europe are routinely full, and staff travel opportunities are almost non-existent because of a shortage of seats and low standby priority. Anything he can do to help is appreciated. When he says to leave it with him, I know he'll do everything he can.

*

After another disturbed night's sleep, my morning starts as it did the day before. I endure a tsunami of emotion and don't fight the tears and grief enveloping me. I'm becoming very concerned with the recurring body response. Is it permanent? Why can't I control it? Is every day going to start like this? I know the flight was a near-death event, but hey, every night carrier landing is like that.

But I'm beginning to accept that this is different – that I really don't have anything in my toolbox of life experience to help me manage this. My body armour has been perforated, and I'm facing a rising feeling of helplessness and frustration. I will fight as best I can and force myself to be patient, hopeful that this wave will eventually pass over me.

*

Ross, Kimberly and I are scheduled to travel back to Sydney as passengers. At Perth Airport, some of the Qantas ground staff recognise me. A small group of them huddle around, offering their concerns for my health and recovery. Naturally, they want to hear a firsthand account of the flight and the emergency. I answer truthfully by saying it was computer-generated but that the exact causes will be found in the investigation. Concern is etched on their faces.

At the departure gate, my crew members and I aren't expecting to see an Airbus A330 connected to the aerobridge, but there it is. I can't believe we're going home in the same type of aircraft as VH-QPA. It appears the Airbus fleet is still flying.

The prospect of getting on this plane is too much for Kimberly. We try to console her, to negotiate a compromise to allay her fears, but it's difficult. Ross steps up and offers to sit next to her. She agrees, and we shuffle our boarding passes to seat them together.

I'm not keen on this travel arrangement. What's going on?

We help Ross and Kimberly get settled, but she's uncomfortable. Ross does an amazing job in calming her down as the flight prepares for departure.

I make eye contact with one of the operating flight attendants, a very bubbly, vibrant young woman. Naturally, she asks what I'm doing in Perth – holidays? I ask if she's seen the news lately about the Airbus emergency two days ago. Her face drops in recognition that we're some of the surviving crew heading back to Sydney.

Meanwhile, the media appears to be taking a keen interest. I haven't looked at any newspapers or television but the story is everywhere. Walking through the terminal, television screens in the departure areas broadcast news programmes showing the images of the plane and analysing the emergency. I only took subliminal notice but this is a big news story.

The aircraft takes off and climbs without incident, and Kimberly calms a bit. I have such enormous sympathy for her and the internal battle she's waging to control herself during this flight.

*

Our flight lands without incident. I'm feeling numb, still in the fog of shock. As we prepare to disembark, Phil tells me that some management pilots want to talk to me in the arrivals hall.

My response is firm. 'No fucking way, Phil. I'm in no state to discuss this now, and I should have a pilot representative for something like this.'

'I understand, Kev,' Phil says. 'John and I will be there as witnesses and representatives. They just want to know what happened.' I think Phil worked at the United Nations in a previous life.

I'm exhausted, emotionally drained and in no state to discuss anything with management. But our discussion is conducted with solemn respect. These two pilots are trained to fly Boeing 747s, so the concept of rogue computers taking control from the pilot and injuring over a hundred passengers and crew is mind-blowing. Computer systems on a Boeing 747 are not as deeply integrated into every aspect of flight as on Airbus planes.

My voice is constricted as I recount the story, and this is also a new experience for me. As I relive the event, my face burns bright red. They know my reputation as an experienced and composed pilot, and it's obvious that my behaviour is drastically out of character.

I end our discussion by saying that I'll assist the company in any way to disseminate this information to the pilot group. I'm also willing to participate in any training development to aid our pilots in dealing with a potential recurrence.

The two men thank me for meeting with them, then let me know that the company will provide me with accommodation in Sydney while I assist in their preliminary investigation.

But I leave the terminal perturbed. The planes are still in the air.

*

It's late afternoon by the time we arrive at company headquarters.

I drop by the office of Adam Lloyd, who has offered his assistance in returning my daughter and her boyfriend to Sydney. He looks tired.

'Hey, Captain Kev!' he exclaims as I enter his small office and slump into a chair.

'Adam, my man,' I answer in the familiar way I have always addressed him during our professional relationship.

'How are you?' Compassion is etched on his brow and in his eyes as we talk.

'I feel like the chariot racer from *Ben Hur* who falls out but still holds on to the reigns.'

We both chuckle at my light-hearted attempt to describe the hell I've been dragged through.

He tells me of the accommodation arrangements: I'll be staying in a hotel in Manly, close to John. Adam then explains his plan for my daughter to fly home: the operating captains of two international flights will ensure seats are available. I'm indebted to them and to Adam – it's a complex humanitarian airlift of a different kind to return my daughter to me.

*

In Manly, I invite my old neighbour, Steve, to share a drink with me. It's closing time in the bar but we convince the bartender to serve us. Over a glass of Tasmanian pinot noir, I relate the story again. Steve isn't a pilot but he knows I'm a straight shooter when I recount flying stories. My changed demeanour is not lost on him. We finish our drink and I say goodnight. I retire to my room. I'm still feeling drained and the bed envelops me. I quickly fall into an exhausted sleep.

But this night's sleep is significant. This time I dream. I'm outnumbered and surrounded by a violent group, bigger and stronger than I am. They pursue me, wanting to hurt me. Despite my best efforts to escape, I'm captured. I'm bound, immobilised, helpless. My antagonists close in, and I know their goal is to maim and kill me.

I wake up yelling. My heart is pounding and my body is flushed. It's the same body response I experienced after the pitch-downs. What the fuck is happening to me?

I sit up and try to analyse my dream. I'm also trying to slow my pounding heartbeat. Is this all just part of the post-adrenalin recovery? It wasn't a flying dream, rather one filled with threat, helplessness and imminent death. Are they connected? I'm reluctant to sleep more; it's early morning, and I'll stay awake rather than return to the hell that's waiting for me in my subconscious.

I'm picking up my daughter at the airport in a few hours, so I might as well get ready.

# 20.

Early the next morning, I'm reunited with my daughter at Sydney Airport.

There's fatigue and relief on my daughter's face; she and her boyfriend have been awake for most of the past twenty-four hours. I don't burden them with my last two days of hell. She recounts the efforts and compassion shown to her by the pilots and cabin crew of the two flights that brought her back to me. Many of these staff members know me and have seen stories about QF72 in the media. It's comforting that such hospitality and respect was given to my daughter and her partner. Crew take care of their own.

*

That evening, someone from Qantas calls. The company's safety department is commencing an investigation into the incident.

I have previous experience with safety and accident investigations. My career in the navy didn't exclusively involve flying: as officers, we were managers first and pilots second.

A navy squadron was a self-contained business unit, so an officer could be responsible for one or two management portfolios within a squadron.

During my second carrier deployment, I served as my squadron's assistant safety officer. It was my portfolio responsibility to ensure the squadron's technical manuals were current, and safety and recurrent qualifications were maintained. I was also actively involved in safety-related investigations, as required.

A naval safety investigation is a fact-finding procedure with the goal of identifying potential deficiencies in procedures, equipment failure or human-factor issues. The results of fleet-wide investigations are shared freely in an effort to improve the knowledge base of the aviation community and to enhance the safe conduct of flight operations.

In this case, the situation should be clear – the best witness to the event is the Airbus itself. The data recorders documented everything the pilots did, along with the confused actions of the flight-control computer systems and subsequent systems failures.

*

At the hotel, I meet up with my daughter and her partner, and we head to dinner at a beachside restaurant. I sense she's aware of my agitation, but this isn't a burden I want to share with her.

'How was your day, Dad?' she asks.

'SNAFU,' I say, which is military speak for 'Situation Normal, All Fucked Up'.

She accepts this with a cute giggle.

The restaurant manager hears our laughter and stops by our table. He notices my Yankee accent and asks where I'm from. Naturally, my answer of 'Sydney' doesn't compute.

'What are you doing here? Holiday?'

'Not really. Did you see the news the other day about the Qantas emergency landing in Western Australia?'

He nods. 'Lots of injuries.'

'You're right. I was the captain of that flight.'

His expression changes immediately. 'Dear Lord! Are you alright? Can I buy you a drink?'

'I'm okay. And yes, thank you. Got any good tequila back there?'

He disappears behind the bar and returns with a bottle and four glasses. We sip and talk, trying to answer his questions about the flight.

My daughter's eyes are wide as I explain the story. I can tell she's also reading my behaviour and taking in how the flight has affected me. This is the first time she has heard the complete account of what occurred on my flight. Naturally, the media would be unaware of these details. She's silent in comprehension and concern. I know I've frightened her.

It's closing time, and we thank this gentleman for his hospitality and service, and I thank him for his compassion.

That night I dream again. It's another threatening nightmare, and the theme is the same as the first one: pursuit, constraint, helplessness, pain and imminent death. I wake up yelling, my heart pounding again. I still can't make sense of this reaction.

I'm stuck in this emotional No Man's Land, confused and frustrated. Night-time is becoming fright-time for me, but it will pass eventually … right?

*

At last I'm free to return to my home on the Gold Coast. Before departing, I inform the company that I'll be taking some time off from flying.

'Take as much time as you need,' I am told.

I don't have a timeframe in my mind, but I know that 'getting back in the saddle', the coping strategy most pilots are taught after experiencing a bad flight, is out of the question. Jumping back in an Airbus would be a completely unprofessional decision on my part. It will take time to recover.

My daughter returns with me, so we can spend a few days together. While we're waiting at the baggage carousel at Brisbane Airport, an off-duty Qantas flight attendant rushes up to embrace me. I struggle to maintain my balance from the force of her hug. We've flown together on previous trips, but I'm surprised at her reaction.

'How'd you know it was me?' I ask nonchalantly, guessing she somehow heard through the cabin-crew grapevine that I was the captain of QF72.

'Because I was at my mum's house, watching morning television and eating my Weet-Bix, and there you were with a megaphone in your hand at Learmonth – and everyone was cheering!'

I haven't seen this video clip, but I remember one of the passengers had a small camera and recorded my briefing in the Learmonth terminal. A major media outlet in Sydney somehow acquired a copy.

The flight attendant asks if I'm okay. I shrug, because I really don't know at this stage, but I tell her I'm happy to be alive and I'll start from there.

My daughter watches all this with amusement. While I retrieve our bags from the carousel, she summarises it aptly and sarcastically with one word: 'Legend.'

I raise one eyebrow, and she smiles. That's my girl.

I spend a couple of quiet days with my daughter, but all too soon it's time for her to return to her life.

My body is still buzzing but now a new situation is developing: a sense of isolation. I don't feel guilty and everyone I was responsible for survived. So why don't I feel like a hero?

# II

# AFTERMATH

# 21.

I retreat to my cave to let the dust settle, try to make sense of my body's chemical rebellion and contemplate the future.

I keep thinking about the decisions I made on the flight and what could have been if the sequence of events had changed slightly. I'm lucky I wasn't physically injured: there was a margin of about five minutes where I wasn't secured while on my way back to the flight deck. If I'd been recalled from my rest break a few moments later and the meal service had finished a few minutes earlier, I would have been in the forward galley – as Pete was when he got injured – checking on my crew before returning to the flight deck for my duty period.

The threat presented to my brain of crashing into the water at the hands of out-of-control computers activated an enormous fight-or-flight response. I'm sure it's the same for a soldier, police officer or first responder – if you're placed in any position where you ask yourself, 'Is my life going to end now?' then your life changes.

I need information about PTSD now, and my first stop is a visit to my doctor. I hope he can shed some light on what is happening to me a week after my wild ride.

A commercial pilot, as part of their annual licence renewal process, must successfully complete a medical examination administered by a designated aviation medical examiner. These examiners are trained as GPs and complete a specialised qualification by the Civil Aviation Safety Authority (CASA) in Australia to administer the annual pilot exams. This process is similar throughout the world as part of a commercial or airline transport pilot's licensing requirements.

Pilot mental health is a recent inclusion in attempts to identify symptoms of depression and alcoholism. It's a pilot's professional responsibility to seek assistance if health concerns could potentially hinder performance. A commercial pilot is directly responsible for the lives of those on every flight, and few occupations shoulder the level of responsibility that a pilot must manage throughout their commercial flying career.

The appointment with my GP starts with my description of the flight. An aviation enthusiast, he's intrigued by the complexity of our failures and the workload we faced. I then recount the body response issues that I experienced during the flight and in the days since. He knows of my military background and agrees that this falls outside the boundaries of my extreme experience. As he describes the symptoms of a 'chemical imbalance' associated with a traumatic event, he writes notes:

1) Mood Swing
2) Irritability
3) Crying Spells
4) Early Morning Waking: 2am – 3am

5) Appetite Disturbance
6) No Motivation
7) Poor Concentration/Memory
8) Low Libido
9) Anxiety
10) Self-Harm

I'm nodding as he reels off these symptoms, then I quietly start ticking off those I've experienced in the past few days. I explain my disrupted sleep, including the nightmares. As we discuss the list of symptoms, my alarm grows. 'Self-Harm'? Could I ever allow myself to do that? Part of me wants to dismiss this discussion as a short-term hiccup, but the other half accepts the precarious and potentially dangerous territory I'm navigating through.

We talk about my libido, and I admit that I feel 'it' has been affected.

'Do you need something for that?' he asks.

'Yes, that would help.'

He scribbles on his prescriptions pad and slides the form over to me as we keep talking. I tell him I'll be taking some time off work, and he starts to fill in a worker's comp form, annotated again with the abbreviation I'm starting to dread: 'PTSD'. The process of documenting my injury has begun.

I collect the various pieces of paper he has filled out for me. Glancing at the prescription for the first time, I see it's for Viagra.

'Hey! What's this for?!' I wave the prescription in front of him.

'You said you needed something for your libido!'

'Not for that! How is this going to help my disrupted sleep?'

He quickly scribbles a script for some mild sleeping pills. 'But do you want to hang on to that prescription too?' he asks sheepishly.

'Actually, Doc, I don't really need any assistance in that department.'

We laugh, and I thank him. But the levity of our banter is replaced by solemn concern as I digest our discussion on the drive home. I stop to fill the prescription for the sleeping pills but don't ever use them. For the first time, a wave of sadness wafts over me.

I'm alarmed by my body response issues and the PTSD label. I know I need time to power down from this event, but I'll get back to work later on … won't I? This internal questioning, in itself, concerns me. I'm working overtime to prevent it from consuming me.

*

With time on my hands, I'm able to explore the flurry of media articles being written about QF72. Initial coverage on 7 and 8 October was inaccurate, citing potential causes for the pitch-downs as autopilot malfunction or clear-air turbulence.

Qantas published their first official statement on the 'Qantas and OneWorld' website on 8 October, under the headline 'Qantas Update on QF72'. It summarised the event and said that the company was cooperating with the Australian Transport Safety Bureau in the investigation. It was also mentioned that an internal corporate investigation was in progress. The Royal Flying Doctor Service and other support agencies were formally recognised and thanked, and the main corporate focus was on the welfare of the surviving passengers and crew. The statement ended by commending the professionalism of the crew in safely handling the emergency.

A day earlier, on 7 October, the day of the event, the ATSB had published its first media release:

> The Australian Transport Safety Bureau (ATSB) has been notified of an incident involving a Qantas Airbus A330 aircraft on a flight from Singapore to Perth. While cruising in level flight, the aircraft experienced a sudden in-flight upset, resulting in injuries to a number of cabin crew and passengers, primarily in the rear of the aircraft.
>
> The crew declared a MAYDAY and diverted the aircraft to Learmonth, near Exmouth in WA, where it landed without further incident.
>
> Early reports indicate that three cabin crew and approximately 30 passengers sustained injuries, including about 15 with serious injuries, namely broken bones and lacerations.
>
> The ATSB has initiated a safety investigation and is making arrangements for investigators to travel to Learmonth as soon as possible.

For the ATSB team of investigators, the QF72 accident presents an enormous challenge in determining the causes of the computer and systems failures that confronted the operating crew and injured so many.

When the ATSB investigation commenced at Learmonth Airport, I spoke with the two investigators who'd been quickly deployed from the Perth office. We secured the data from the flight-data recorder and the cockpit voice recorder on shutdown in accordance with instructions provided by the maintenance unit. ATSB personnel supervised the removal of the recorders from QPA, and they were sent to the ATSB offices in Canberra for inspection, arriving on 8 October. The downloaded data was provided to the investigating team on 9 October.

*

On 14 October 2008, the ATSB team present their initial findings at a media conference in Canberra. Examination of the recorded flight data from the aircraft has revealed significant failures of the automation, specifically that the air-data processing portion of ADIRU 1 produced 'spiked values' of data that were responsible for the false stall and overspeed warnings, the loss of the captain's instrumentation and the loss of other systems.

This falsely generated data was responsible for the pitch-downs ordered by the flight-control computers and the subsequent faults with the electronic flight-control systems.

The ATSB team comment on the actions of the flight crew: 'The crew's timely response led to the recovery of the aircraft's trajectory within seconds.'

On the same day, 14 October, VH-QPA is flown out of Learmonth to Sydney. It's a positioning flight by pilots from the Qantas Technical Section and carries a contingent of engineering specialists providing technical support. An extensive refit of the avionics and a clean-up of the interior has been completed in the seven days following the emergency landing.

Many of the electronic components have been removed for testing as part of the ongoing investigation, and Airbus has provided technical assistance to Qantas in preparing the aircraft for the flight. Extensive testing of the aircraft's wiring and the operation of the flight-control system is completed before the flight is conducted.

On 15 October, Airbus issue the first of several operational engineering bulletins to address issues with abnormal operation and faults associated with the air-data computers. This is in response to the ongoing ATSB investigation, and is formulated after the recorded data from the aircraft has been downloaded and analysed. The enhanced procedure directs pilots to turn off both the inertial and air data portions of the ADIRU if any faults are displayed on the pilot's instrumentation.

The same day, *The Sydney Morning Herald* publishes an article entitled 'Ex-fighter pilot saved QF72'. A colleague from our pilots' association has been interviewed. The article focuses on my US Navy flying experience and my documented coolness under pressure in previous emergencies; there's also comment on the training of Qantas pilots and how it would have assisted in a successful outcome. The article concludes that the malfunctions the crew faced were outside the scope of normal or expected operation, underlining the fact that Qantas pilots are held in high regard around the world and are known as safe and reliable pilots.

# 22.

The Australian Transport Safety Bureau team schedules our formal interview for 21 October in their Brisbane office as part of their investigation. All three pilots are in attendance, as well as Captain John Killingback, acting as our pilot representative.

There are three ATSB investigators at my interview. The atmosphere is cordial and professional, and I'm not nervous – they're analysing and reporting on an accident that's way outside the boundaries of normality.

The interview is recorded. After providing a written summary, I'm asked to recount the events of the flight and my actions in responding to the pitch-downs and emergency landing. They listen quietly, making notes as I progress through the sequence of events for the fifty minutes we were airborne. They clearly understand the severity and magnitude of the workload the pilots were faced with, and the confusing information and lack of communication that the computers were throwing at us. I describe my bewilderment when my stick inputs couldn't counter the computer's nose-down manoeuvre. I express my frustration that the aircraft computer systems didn't supply any information to justify the activation of the protection modes or the data being sampled to generate the computer's violent response. Airbus has already established that protections modes were activated, causing the dives.

I let them know why I did a control check at altitude and made a high-energy approach to landing; I didn't trust any of the information or the automation that the aircraft was providing. In my opinion, I tell them, the computer-commanded dives and multiple systems faults generated a confusing, distracting and dangerous cockpit environment.

The investigators ask me if I want to listen to the cockpit voice recorder. This captures everything said or heard in the cockpit from the time the engines are started until they're shut down at the end of the flight.

I'm still emotionally raw – I haven't been sleeping well and don't want to stress myself further. 'I don't want to listen to it now, but can you keep the invitation open for later?' They agree.

'Do you remember what you said when your second officer asked if you wanted to reset PRIM 3 after the second pitch-down?' one of the investigators asks.

'Yeah. "Don't fuck with it."'

'That was my favourite part,' he says.

'I think I used the f-word a few times that day,' I reply wryly.

After the interview, the investigators ask me to sign a confidentiality statement in regard to the ongoing investigation, and I'm briefed on the investigation process. The aircraft's data and voice recorders are intact, and the investigators have three eyewitness accounts from the pilots.

'How long will it take to complete?' I ask.

'Unsure at this stage, but we'll keep you updated.'

*Should only take a few months to complete the investigation*, I think to myself.

I wait outside as Pete and Ross complete their interviews. When they emerge, separately, their faces are troubled.

'You didn't listen to the voice recorder, did you?' I ask.

They both nod. Listening to the tape has triggered some very bad memories, still raw and unresolved in our minds. This affirms my decision to wait longer before reliving the event through the recording.

*

A week later, I meet with the crew of QF72 in Sydney. It's the first time we've seen each other since the flight three weeks earlier.

Their injuries are still visible: facial bruises, injured limbs, stiff necks and shoulders. And my heart breaks as Fuzzy hobbles in on crutches to greet his fellow crew. He moves awkwardly and looks in distress, but maintains a dignified air of courage and humility. My eyes tear up as we shake hands and embrace. How could this have happened to him? To us?

One by one, my brave crew recount their experience on that horrible day. Their eyes lock onto mine as they describe their reactions to the pitch-downs, how they were injured, their fears and confusion, their actions during and after the violent manoeuvres, and their dedication to their passengers and to each other. My heart is pounding again – not only in sympathy for their injuries and their trauma, but also in admiration and pride.

For Rory it is his first opportunity to release his pent-up emotions and seek support from his fellow crew members. He's the flight attendant who ran through his section – with a fractured foot, it turns out – checking on his passengers while Lisa and others yelled at him to be seated. His emotional reconstruction is powerful and heartbreaking. He details how he was thrown into the ceiling while moving through the cabin after the meal service; he came down hard on his foot and was dazed, thinking the aircraft was in turbulence.

After the second pitch-down he disregarded his injury as he moved rapidly through his area, instructing passengers to fasten seatbelts and checking on the infants in his section before securing himself in a seat. On landing, he quickly retrieved a first-aid kit from its stowage.

As he moved through his cabin area, an elderly passenger with a bloodstained pillow held tightly to his head stopped him for assistance. After the pillow was removed, Rory was shocked to see a jagged gash along the man's scalp line, part of the skull exposed. This man hadn't fastened his seatbelt and had been propelled through the plastic above his seat. He grew concerned at Rory's shocked expression and asked if his injury was very bad.

'No, mate, no, you're gonna be okay,' Rory replied, then replaced the pillow firmly back on to the wound. As Rory recalls this event, he begins to cry, all the while staring into my eyes, which are also stinging with tears.

I have to keep looking down, to feign rubbing an itchy eye as I fight back my tears. I realise that the crew never really talked to me about their individual actions until now. I knew they were in shock and traumatised in Learmonth and at the time I didn't want to make them relive the event.

When it's my turn to talk, I remind the crew they'd done an incredible job. Faced with mass casualties, they used innovation to deal with the unexpected emergency. They communicated with each other and with the passengers to clear the aisles and prepare for the contingency of an evacuation. Finally, they supported each other and their passengers after we landed until the medical response was completed and all the passengers were moved into the terminal. They're response couldn't be faulted.

But sitting with them now, I could see it would take more than words for this group of remarkable individuals to recover from the experience. It had left a permanent mark.

*

My return-to-work coordinator, Adam Lloyd, helps me navigate the complexities of extended leave and worker's comp intricacies.

Meanwhile, Pete and Ross are keen to return to flying and need to requalify, but are uneasy about flying the Airbus fleet, with the causes the subject of an ongoing investigation. In my opinion, the get-back-in-the-saddle approach after such a unique traumatic event can be a challenge.

*

On 14 November 2008, the Australian Transport Safety Bureau releases a preliminary report on QF72. The chief ATSB investigator informs the three pilots the week prior to the publication, also letting us know a draft copy will be provided to us before the public disclosure. I appreciate that they're keeping us informed.

Examination of the data recorded by the aircraft's black boxes has identified abnormal data production and output from ADIRU 1. The investigators are now able to match the failures with a time marker. They can also identify the failure sequence caused by the erroneous data generated by the air-data computer and its domino effect on the computers and automated systems.

Of significance is the data recorded from ADIRU 1 regarding angle of attack. This raw value is captured by each of the three metal probes on the sides of the aircraft below the cockpit. Simplistically, it's a measure of the aircraft's body angle relative to the air mass, but really, it's an angular measurement of the aircraft's wing relative to the air mass. In cruise, an aircraft has an angle-of-attack value of +2 degrees, and this was the recorded value before the autopilot disconnected at 1240 (Perth time). The angle-of-attack values recorded after that were the 'spikes' mentioned previously, and their peak values were +50.6 degrees.

Angle of attack is a critical value for a fly-by-wire aircraft. The computers responsible for keeping the aircraft safe need to know where the aircraft is inside the air mass it's travelling through. The air-data values generated inside the three air-data computers provide this situational awareness to the flight-control computers. Each air-data computer produces and transmits its angle-of-attack value sixteen times per second to each of the flight-control computers, and that data is recorded each second by the aircraft's black-box recorders. A lot can go wrong within those one-second recorded points.

The recorded angle-of-attack spikes of +50.6 degrees are huge values. Even high-performance military aircraft don't operate in these regimes, except under extreme manoeuvring, and there has to be something seriously wrong with ADIRU 1 for these values to be generated internally. A value of around +2 degrees would have been measured by the aircraft's angle-of-attack probes and sent to the air-data computers. Somehow, ADIRU 1 took that routine value and magnified it to +50.6 degrees before distributing it to the three flight-control computers. This extreme value exceeds the wing's stall angle: the angle where the air over the wing separates and the wing's lifting force is lost.

The ATSB has determined that these spikes in angle-of-attack generated the stall warnings – of which forty-two were recorded, but there could have been more due to the black box's data-capture recording limits of one second. There were more, I can confirm that.

Commercial pilots aren't able to monitor angle-of-attack values through any available instrumentation, so we're oblivious to the values going into and coming out of the ADIRUs. But navy pilots are trained to use angle of attack to set various performance values. I've long held the opinion that angle of attack is a useful performance tool in commercial aviation; however, it's only being embraced in newer-generation commercial transport designs.

Had I been alerted, through appropriate instrumentation, that the aircraft was generating random angle-of-attack values of +50 degrees, perhaps I would have been able to address that scenario more effectively.

The stall and overspeed warnings generated by corrupted data from ADIRU 1, though found to be spurious, were continuous and unable to be silenced. These warnings are serious and require immediate pilot action if they're valid, so their activation was an enormous distraction to the flight crew while we attempted to control the aircraft, gather information and generate a landing strategy.

The preliminary report also comments on the potential for electromagnetic interference, from either personal electronic devices used by the passengers or from the Naval Communication Centre Harold E. Holt. This is spooky stuff.

It's painfully obvious to me that these many failures are extremely complex and outside the realm of the Airbus training course or technical information provided at the time. How can an aircraft be certified as safe if it its design allows this type of dangerous behaviour? How can the ADIRU corrupt and produce extreme data from benign values, and how can the flight computers use that data aggressively?

The preliminary report confirms to me that this accident was a science-fiction scenario that became a reality: automation gone haywire. The humans, as the last line of defence, were ill-equipped to stop it. There is no disconnect, no fail-safe switch, installed to override this type of computer malfunction from occurring on an Airbus.

I use the film *2001: A Space Odyssey* to understand and accept what happened on the flight. I was Dave, negotiating with the sentient computer HAL, to 'open the pod bay doors'. My commands to arrest the dives were overruled: 'I'm sorry, Dave. I'm afraid I can't do that.' The aircraft wouldn't let me stop it from protecting itself.

# 23.

Another Airbus is lost in the month following QF72.

On 27 November 2008, an Airbus A320 crashes near Perpignan, France, in the course of completing a test flight. Air New Zealand had leased the aircraft to XL Airways Germany, and it was being returned. The pilots somehow lose control of the aircraft on approach. All seven crew members perish.

A photo shows the vertical tail submerged in shallow water, emblazoned with the Air New Zealand logo, as news of the tragedy is reported. I wonder: Could this accident have similarities to QF72? How could the Airbus design, with its infallible protections, allow this type of loss-of-control event to occur? By now, I'm seriously questioning the safety of the Airbus design, and it troubles me greatly that an Airbus has gone down so quickly after my own event.

The interim report, released in February 2009, points to pilot error as the cause of the crash. I examine the data and the graphs at the end of report, noticing that both angle-of-attack sensors appeared to have failed early in the flight, but no mention of this is made in the interim report. Even when the aircraft had departed controlled flight, it showed a constant low angle-of-attack value. I'm surprised and baffled by this omission, but most reports from this European states investigative agency are quick to point the finger at pilots in any accidents involving Airbus.

The following year, the final report pinpoints water intrusion into the angle-of-attack mechanisms after the aircraft was washed and the sensor area hadn't been covered correctly. This water froze early in the flight at higher altitude, also freezing the angle of attack at a constant value, which was then sent to the flight-control computers. Sadly, there was an element of pilot error involved, but the frozen angle of attack played a part.

Media around the world are quick to question if this accident has similarities to QF72. The resounding answer from 'aviation experts' is 'no'. But I believe they're similar. Air data was 'corrupted' through the frozen angle-of-attack probes, and the aircraft used that corrupted data. The electronic flight-control computers didn't alert the pilots to an anomaly with the angle-of-attack value. The management computers responsible for displaying airspeed and altitude to the pilots are also responsible for displaying the angle-of-attack limit on the pilot's primary display. The frozen angle-of-attack value subtly modified the limit value displayed on the pilot's airspeed indication.

The pilots made a mistake; they rushed their protection check at low altitude, and they didn't calculate the indicated speed at which this protection should activate in accordance with their checklist procedures. This combination of corrupted air data and pilot confusion generated a loss-of-control event.

In the course of their attempts to recover the aircraft, the flight computers subtly reverted to the basic level of flight protection, called Direct Law. This allowed the pilots to stall the aircraft and presented a complicated recovery situation that they weren't able to accomplish.

Analysts have started to quote a new term for pilot error: 'startle factor'. In my opinion, this factor is subliminally built into the Airbus design. We were certainly served a huge dose of it.

The sensor probes are subjected to the extremes of the air mass in which they're required to operate. They can be blocked, frozen or damaged, but they will still send a raw signal to the ADIRU, meaning a corrupted value of air data will be displayed to the pilots. It can be a complicated task for a pilot to effectively recognise unreliable speed, angle of attack or temperature, given that the aircraft might be experiencing bad weather, turbulence or systems failures.

If the Airbus is so smart, why doesn't the automation provide this recognition function? Why depend on the pilots to recognise this degradation in air data and potentially get it wrong? The QF72 crew was presented with a similar scenario, but we survived. It's a scenario that becomes repetitive in the accidents and incidents on Airbus aircraft in the years to come.

*

By now, VH-QPA has been repaired and tested, and is now ready to be reintroduced into normal operations.

Then another automation event occurs.

On 27 December 2008, a Qantas Airbus A330-303 aircraft, registered VH-QPG, is flying from Perth to Singapore as QF71. While the aircraft is in cruise at 36,000 feet, the autopilot disconnects and the crew receives an alert of a failure in ADIRU 1. This computer is the same model as that involved on QF72.

Previously, Airbus had produced a new procedure to address non-normal ADIRU operation, which didn't exist prior to QF72. It involves procedures to isolate either the air data or navigation portions of that computer, to eventually shutting down the computer completely for a given set of circumstances. It was in the third iteration at this time and would change again as more testing and operational feedback occurred.

The crew of QF71 are prepared, as most Airbus crews are at this time, with Version 3 of the procedure pre-positioned on the instrument glareshield, and they quickly execute the necessary steps before the aircraft's computers can use the corrupted data. Their prompt actions prevent computer-generated manoeuvres, but they keep receiving warning and caution messages that constantly scroll on the display and can't be silenced. The crew elects to return to Perth and an uneventful overweight landing is conducted.

After the QF71 event, the operational procedure is further modified to Version 4 in January 2009.

Sitting on the sidelines, I quietly pat myself on the back for taking time off from work while all these procedural changes from the manufacturer are modified and published. That could have been me operating as the captain of this flight. What happened on QF71 shows there are still issues with the Northrop Grumman ADIRU and the Airbus software.

*

In February 2009, the chief ATSB investigator notifies the QF72 pilots of the impending publication of an interim report. Prior to its release, we're provided with a copy to comment on. The report builds on the foundations of the preliminary report, doing more to analyse the technical complexity of the failures of the Airbus electronic flight-control system and its operational partnership with the Northrop Grumman air-data computers.

On 6 March 2009, the ATSB publishes the interim factual report. Fifty-three pages long, it features updates on the number and types of injuries, seatbelt and cabin safety issues, and ongoing ADIRU testing. Of significance, there are more answers on why the aircraft computers behaved so aggressively. Through the partnership with Airbus, the data spikes have been re-created by their massive systems computer, nicknamed the 'Iron Bird'. According to Airbus, the Iron Bird

> is an engineering tool used to design, integrate, optimise and validate vital aircraft systems. Since all aircraft systems are controlled from the flight deck, the Iron Bird needs a cockpit for its control. Three Fixed Based Simulators (FBS) are used along with a mobile visual system. After certification and when the aircraft is in revenue service, the Iron Bird is used for further development of the aircraft systems as well as a test bench to trace anomalies that may show up with components or systems.

Airbus successfully simulated the data spikes and found that the installed software in their flight-control computers, designed to filter these spikes, was flawed. The software algorithm allowed the flight computers to accept the extreme angle-of-attack values as real, and therefore ordered the two pitch-downs. This discovery identifies a rare and unforeseen scenario that wasn't considered in its design or during its certification testing. This scenario lay dormant in the spaghetti of all the computer software until the perfect storm of QF72 generated the opportunity for it to fail.

I begin to wonder how many more flawed algorithms are potentially lying in wait within the software.

The report then describes what the computers commanded to cause the dives and injuries. The extreme angle-of-attack and other air-data values data produced by ADIRU 1 fooled the flight computers into accepting the aircraft was in danger. They simultaneously activated two protections to keep the aircraft safe.

Unlike the pilot, the flight computers have direct control over the aircraft's elevator control. This is the control surface on the tail that moves the aircraft up or down; the computers can move it instantaneously and precisely if required. A pilot must request this movement through either the sidestick or the autopilot, and extreme requests are filtered by the software to maintain the aircraft in its safety box of operation.

The two protections ordered by the computers each have a value of tail movement, producing a downward flight path. The first protection, 'High Angle-of-Attack Protection', is worth four degrees of elevator movement. The second protection,

'Anti Pitch-Up Compensation', is worth six degrees of elevator movement. It's simple arithmetic: 4 + 6 = 10. That's close to the maximum downward tail movement of 14 degrees. On QF72, the computers combined the two protections to generate 10 degrees of tail movement. The aircraft's attitude was +2 degrees above the horizon while in cruise. The computers' command pushed the nose down to -8.4 degrees below the horizon – within only two seconds. At 37,000 feet and at the fast cruise speed we were flying, this pitch rate generated an enormous downward force.

The pilots, the last line of defence, were preordained by the Airbus design to be placed in a powerless position to stop those erroneous modes from activating without an instinctive control or override built into the design. When protections are active, the pilot's sidestick commands are ignored. The nose movement looks benign on the ATSB website animation, but the elevator was moved aggressively to near maximum deflection within those two seconds. A human pilot can't physically perform this precise pitch-down manoeuvre as effectively and violently as the computers can.

I still can't understand how the software allowed two protections to be added together and generate the same movement at this high altitude as it would at a lower one.

The report classifies what happened on QF72 as an 'accident'. The a-word: one that airline operators dread and despise. Of course, Qantas is known for its impeccable safety record, globally – even in Hollywood. 'Qantas never had a crash,' as Dustin Hoffman's character proclaims in *Rain Man*.

The ATSB explains the reason for this accident classification, a result of the many serious injuries experienced on the flight.

Qantas, like many other airlines, has experienced some serious events in its recent operating history. However, if there are no serious injuries and if the airline can repair the damage and return the aircraft to operation – as it did after the QF30 and QF32 events – the event is downgraded from an 'accident' to an 'incident'. This preserves an airline's 'accident-free' record.

VH-QPA sustained substantial damage to the interior of the aircraft, but it was repaired. Sadly, however, there would be no quick fix for those seriously injured on QF72.

This accident had nothing to do with the airline, and yet it would be stuck with an accident in its safety history.

The ATSB interim report has identified the root causes of the pitch-downs and the errant generation of air data, but all the 'whys' are yet to be determined. After reading this description of the pitch-downs, and learning that the flawed software is dormant, I again question – as do my colleagues – why the Airbus fleet hasn't been grounded.

The investigation has a long way to go. It's April 2009, and I have some big decisions to make.

The name of a song by the Clash keeps going through my mind as I contemplate my future: 'Should I stay or should I go?'

# 24.

In most airlines there's a seniority system. On your first day at work, you're assigned a number. This determines which aircraft type you fly, along with leave periods and the base city you'll fly from. As you progress in an airline career, your opportunities to advance and fly the long-range aircraft, with more salary and overtime pay rates. Your position relative to your colleagues' remains constant, determined by your seniority number.

Being a captain flying the Airbus A330 is the highest position my number allows. The other non-Airbus fleets are either being deactivated (767) or aren't accepting new positions (747-400). Nevertheless, in May 2009, I decide to return to work.

I've thought seriously about stopping permanently, but I won't let go of my lifelong passion without a fight. The fight, I've realised, is with myself. I don't think anyone will think less of me if I don't return to flying. Many of my colleagues appreciate the severity of my ordeal, along with the price my crew and I have paid to 'save the day'. Whenever I'm asked about what happened, I struggle to relate the challenges we faced – my voice cracks, my face flushes, and my speech becomes louder and more animated as I summarise the flight and the frustrations afterwards.

In fact, eighteen months pass before I can tell the whole story without my voice cracking. This confirms to me that I've sustained some permanent damage, but I won't give in. I need to look at myself in the mirror and into my daughter's eyes, and proclaim that I did not quit. I won't hide my damage, nor will I let it stop me from trying to reclaim my life and my career.

Vince Lombardi is a famous American football coach for the Green Bay Packers, who's widely recognised for his inspirational quotes. 'It's not whether you get knocked down, it's whether you get up,' he says.

I will get back up. I will face my demons and the A330; after all, everything has been fixed, right? And my seniority number dictates that I don't have any other choice if I want to keep working for Qantas.

My return-to-work coordinator, Adam, is relieved when I officially inform him of my decision. I've been off work for eight months, and the limits of worker's compensation payments at the highest rate are approaching the twelve-month time limit.

Then, less than a week after I state my desire to return to work, Air France 447, an Airbus A330, is lost over the southern Atlantic without a trace. The date is 1 June 2009.

Adam immediately calls me to discuss this latest Airbus tragedy. 'How do you feel about this accident?' he asks.

'What happened?'

'They don't know. It disappeared without any communication. The authorities don't even know where the wreckage is.'

I already know all of this, as I've been following the story intensely. Another Airbus is down … why?

'Well,' I say, 'I hope it was a bomb and not a problem with the flight-control computers like on QF72.'

Air France 447 disappeared over the deepest part of the Atlantic Ocean. The event occurred at night and without any communication from the crew of any difficulties. In my mind, it fits the profile of a terrorist in-flight bomb attack, not a computer fault. I have experience in this area: after the 9/11 attacks, I worked on secondment to the Qantas Security Department as an aviation security specialist.

'How do you feel about returning to work after this?' Adam asks.

'What difference does this make?' I reply. 'I'll return as discussed.'

The wheels are in motion, but it will be up to me to tick the many boxes required to satisfy all the parties involved with my employment and licensing. The admin road back to the flight deck isn't as clear-cut as I expected; it seems they've had to re-invent the wheel to cater for my unique circumstances.

The PTSD condition attached to my Airline Transport Pilot License means that the Aviation Medical Branch of the Civil Aviation Safety Authority has placed me under 'special audit'. I'm considered a 'complex case' that requires a review and assessment by a panel of aviation medical specialists before my medical qualification is renewed and I'm cleared to undertake a requalification programme at Qantas. This vetting process is very frustrating, but it makes sense. When a pilot who's been diagnosed with PTSD returns to the cockpit, the situation is full of uncertainty. But while my fight-or-flight body responses are referred to as 'panic attacks' in some circles, I'm not panicking and have taken exception to this label being attached to my situation.

Once these requirements are met, I receive my airline transport pilot licence and aviation medical, then commence my requalification in late June 2009. I'm certified as fit to fly, and that's how I'm assessed.

My hope is that once I'm back in command I'll desensitise and my body responses will diminish. What else can I do but try? There's no instruction manual for PTSD, so it's up to me to write one specifically for me. This will be a rocky road of trial and error, and no amount of preparation can prepare me for the corporate environment I'm returning to.

One of my first duties back at Qantas is to complete the emergency procedures course to renew competency in all the cabin safety equipment.

One half of the day is a pilots-only discussion, which involves analysing a mishap from another airline or from an internal event. The moderator asks me if I'll tell the assembled group about QF72. Red-faced and with halting speech, I describe the event and its effects on me, particularly my body's fight-or-flight response.

My next duty is a re-familiarisation exercise in the flight simulator, supervised by a training captain. 'This is a freebie for you to get reacquainted with the A330, and to explore computer faults and the new operational procedure developed by Airbus,' he explains. But I already know the QF72 failures can't be replicated in a flight simulator, so I simply nod.

In the sim, I take the left seat and my instructor takes the right. After a routine review of cockpit preparation and starting the engine, we're in the simulated air.

Even with eight months away from the flight deck, I find it easy to manage the A330. When everything works, it isn't a difficult task; that's the one benefit of its design. Failures are introduced to simulate, in my instructor's mind, the QF72 failures. I handle them effortlessly because they're singular, not multiple, while the warning system works as it was designed to, allowing pilot interaction and reconfiguration.

I decide to demonstrate the pitch rate from +2 degrees to -8.4 degrees that occurred on QF72. I don't want to break the simulator, so I ask him to turn the 'motion' off, cutting power to the hydraulic pistons that move the simulator as it emulates the accelerations and decelerations experienced in normal flight. Once he secures the hydraulics, I slam the stick forward then back, trying to match the computer's pitch rate – but the software in the flight-control system doesn't allow me to move the tail as precisely as it can.

I point out that there's at least a -0.8g acceleration generated at the aircraft's recording point, but it was a much higher force in the back galley. 'We're lucky,' I say, 'that no one was killed as their heads and bodies slammed into the ceiling and through the plastic.'

My instructor's 'book knowledge' is exceptional. However, the books don't cover what happened to us.

After we're finished in the simulator, he asks for my feedback on the session.

'Acquaint yourself with the ATSB report and discuss its contents at every simulator session you conduct,' I respond.

My training package starts the next day. Back in 2004, my initial training lasted eight weeks; this time it will take me three weeks to complete the package. My performance is assessed unconditionally, without consideration of the traumatic experience of QF72. I believe this is the correct approach to my requalification. I'm as focused and determined to succeed as I've ever been in my flying career.

# 25.

One month after starting my training package, I'm cleared for supervised flying in the A330. My first trip is to Narita, Japan. A training captain is there to supervise my performance.

I've decided to manage my flights in a different way from now on. The mass casualties on QF72 are at the forefront of my mind, and because I never want to revisit that scenario, I've changed my pre-departure routine to cater for my newfound precautionary approach.

As always, I make the time to walk through the cabin, introducing myself to the crew. They're my team, so before we depart I want them to know who their captain is, and I want to make sure the lines of communication are open. In this age of secure flight decks, accurate and open communication from the cabin crew is essential in managing the many challenges that can develop during a long-haul flight. I want to cement trust between captain and crew.

After saying hello, I pass on the flight-time details and any anticipated turbulence areas that might affect their cabin service. Now I look each crew member in the eye and state that 'no one gets hurt today'. I encourage them to contact the pilots if things get rough in the cabin. For some reason – and I've seen it firsthand – pilots can hesitate to put the seatbelt sign on during turbulence. It feels very different at the front of the plane compared to the rear of the cabin. The lightweight materials used in the construction of the aircraft's structure generate more strength-for-weight, but also there's more flex throughout the tube of the airframe. In turbulence, this flex is very apparent; you can see it if you're in a position to look aft and see the tail swaying from side to side.

After my experience on QF72, I will never hesitate to put on the seatbelt sign and secure my passengers and crew.

Back on the flight deck, I'm engrossed in pre-departure checks and take-off performance calculations. Over the years, I've developed disciplined habits and I tap into them as I methodically work through the phases of the departure sequence.

Soon we're at the holding point for the runway, ready for take-off. My heart rate is increasing, a bit more elevated than usual. Take-off and landing are the most critical phases of any flight – the proximity to the ground and the restrictions of landing gear and flaps make any abnormal situation at low altitude more difficult. I don't feel nervous or excited, but my body is generating more chemicals than in the past for this rather routine phase of my flight. I compartmentalise that for later analysis and concentrate on the heavyweight departure.

Safely airborne, I engage an autopilot and transition into the more mundane monitoring of navigation and fuel usage.

Internally, I'm on edge; externally, I'm as cool as the other side of the pillow.

'How does it feel?' my training captain asks, with a wry smile.

I ask myself the same question. Too early to tell, but it feels like I haven't been off-duty for this eight-month period.

'These planes fly themselves, don't they?' I respond.

As we climb, the air mass is smooth and conditions are safe for the crew to commence their duties. I monitor the CSM directing passengers to 'keep seatbelts fastened while seated' on the PA; not much more I can do on that topic as it is also part of my passenger greeting. But I know we'll encounter thunderstorm activity in four hours as we overfly Papua New Guinea.

I'm not relaxed. The elation and joy I felt before are missing. I'm a circus bear, jumping through hoops under the supervision of my training captain. But if he's satisfied, then I'll be on track to getting back into command.

As we approach New Guinea, I'm manipulating the aircraft's weather radar controls. On the horizon, billowing clouds flash from the energy of cumulonimbus formations growing over the mountains, while lightning bolts arc down from some of the more active cells. I call the back galley to let them know the seatbelt sign will be coming on and they must secure their areas in anticipation of turbulence.

My weather radar display is like an oil painting. Green, yellow, red and magenta blobs depict the varying moisture content of the cells and let me know about the areas of turbulence associated with the thunderstorms: green is light moisture and turbulence, while magenta is severe turbulence. After reading about the XL

Airways and Air France 447 accidents, I've become even more conservative in managing my flight path to avoid precipitation and unpredictable temperature changes in the vicinity of these high-energy cloud formations.

I ask the first officer to illuminate the seatbelt sign and direct all passengers and crew to fasten seatbelts. As he makes his call, I scan my navigation display and plot a course around the biggest cells. I try to remain upwind of the thunderstorms but that won't be possible tonight. A big moon is helping to illuminate the cells and their boundaries, as well as the lightning flashes silhouetting their vertical development. The complete vertical structure of a thunderstorm can be hidden from our weather radar. I'm using every clue to assist my deviations off our planned track to avoid the weather.

The band of adverse weather is at least 100 miles across, and I'll just have to weave my way through as best I can. I decelerate to turbulence-penetration speed and turn on the engine's anti-ice protection. As we enter the clouds, the aircraft reacts to changes in moisture, temperature and wind direction.

All my preparations to protect my passengers, crew and airframe are complete – now I have to work the radar as I find a way through the maze of cloud.

The windshields turn opaque as we enter the clouds, and the flash of the red anti-collision light and white wingtip strobes produce a disco light effect as we dodge the cells. The plane starts moving and feels like a car driving over cobblestones.

I'm scanning my instruments as I double- and triple-check the path of my avoidance plan. Rain is pounding on the fuselage and windscreens. The temperatures check okay, but I know that thunderstorms can change the temperature rapidly and severely. I'm not enjoying this rough ride very much; it's like sitting in the dentist's chair without novocaine while the drill whines and grinds into your teeth.

Suddenly, the plane starts to climb. The vertical speed indicator confirms my seat-of-the-pants sensation. A second later, the plane descends sharply. The autopilot is pushing it back towards its commanded altitude, and I feel light in my seat from the slight negative force. Instinctively, I grab the instrument glareshield with my right hand and move my left further onto my control sidestick. My stomach spasms as if someone has reached in and squeezed it. Annoyingly, my heart rate increases. I'm going into hyper-drive as I scan my instrumentation for any threat but, in an instant, the aircraft stabilises.

I take a deep breath and glance over at my first officer, calmly lounging in his seat with a slightly miffed expression as he observes my response to this little bump. I apologise for my abrupt reaction to this normal aircraft movement in turbulence.

This is my first introduction to the new me in the cockpit.

We exit the storm, revert to the routine of cruising in clear air and re-establish our planned route towards Narita Airport. I'm taking more deep breaths as covertly as possible, and the significance of my body response to the minor turbulence isn't lost on me. I'd never reacted like this, even in the most severe turbulence, prior to QF72. Something isn't right.

*

The old me never experienced panic or startle, not even when I was operating around an aircraft carrier at sea. I recall an incident when I was deployed on my first cruise aboard USS *America* as a 'nugget' pilot flying the F14. We were somewhere in the Arabian Sea, maintaining station and keeping the world safe. I was learning my trade as a naval aviator and I was learning fast.

I was on the wing of my flight lead, returning to the ship from a routine combat air patrol mission. It was daytime and the sea was calm, perfect conditions for a day trap aboard *America*.

After breaking into the landing pattern, I went through the routine drill of configuring for landing: wings swept forward, landing gear and flaps extended, hook down and harness locked. I decelerated to my approach speed and was trimming the tail for the proper nose position for landing.

I advanced the power but only one throttle would move. I tried more force to move the uncooperative throttle forward but no go. My left engine was stuck at idle power. I'd need more time to prepare for this approach and didn't want to disrupt the landing recovery of the other aircraft lining up behind me. I needed to inform the Air Boss of my situation.

'Boss, 110.'

'Go ahead,' came the gruff response. (Are all air bosses perpetually cranky?)

'110. Stuck throttle on the left engine. Request delta and straight in approach to recover last.'

'Delta' meant I would orbit above the normal landing pattern altitude while the rest of the aircraft landed. It would also give me some time to troubleshoot my problem and prepare for a single-engine approach. Landing last would ensure minimal delays for the recovery if I encountered any further complications on landing. There were no land-based divert fields near the carrier.

'Roger 110. Delta.'

With my experienced RIO, we went through the procedures for my approach and landing. I would need to fly a straight-in approach for our degraded condition and I maintained orbit behind the ship.

As the recovery progressed my situation started to deteriorate. An amber 'left oil pressure' caution light illuminated on the system panel. We quickly checked our emergency flip pad procedures. We decided to keep it running as our turn for landing was imminent.

I informed the Boss of our situation and we were cleared for approach and landing.

A single engine approach is a big deal in the F14. The engines are over 10 feet apart and the yaw generated requires careful use of the rudders to balance the aircraft. Also, precisely flying the single engine 'angle of attack' prevents the jet from torque rolling around the inoperative engine. We routinely practised this critical failure on land, but accidents still occurred at sea when these parameters were not controlled effectively.

Calmly and methodically, I prepared for my approach. The deck was clear and the meatball in sight as I lined up on the ship. I was halfway through this first deployment and I was a seasoned carrier pilot by now.

I manipulated my power precisely to keep the meatball centred and each power change required a rudder input to keep the aircraft balanced and aligned with the ship's landing area. Small, coordinated flicks of the wings kept me on the centerline.

My RIO called the ball and added 'stuck throttle approach' to the ever-vigilant LSOs watching my jet like hawks.

'Roger ball,' came the sharp response of the controlling LSO. And then he added: 'Secure your dump.' He was seeing something I wasn't!

The 'dump' is the fuel dump valve that we routinely used to dump fuel out of our tanks to adjust our fuel quantity and weight for the arrested landing.

I wasn't dumping fuel.

I was on full alert, flying my crippled F14 towards the welcoming arresting wires. As I approached the back of the ship, I got another caution light but I was too close to landing to worry about it.

I felt the forceful decelerating grab of the arresting wire as I landed and advanced full power on my good engine.

'Secure your left engine!' the LSO screamed.

Instinctively I reached up and pulled the 'emergency fuel shut-off' handle for the left engine. My quick scan of the caution lights showed an amber 'left oil temperature' brightly illuminated.

The flight deck fire crews quickly surrounded my Tomcat and gave me the cut signal to shutdown my other engine. Were we on fire?

I announced 'emergency egress' to my back-seater. We quickly unstrapped and ran clear of our smoking Tomcat.

I was in a huddle with my squadron maintenance crew when the LSOs appeared to debrief my landing. My left engine was smoking so badly they thought I was dumping fuel.

I watched my landing soon after on the closed circuit TV in our ready room. It looked like Thomas the Tank Engine was hitching a ride on the back of my F14. Thick grey smoke billowed from my left engine as I flew my approach.

My jet was taken down to the hanger bay for an extensive repair. An oil pipe had failed and the resulting heat had fused the mechanical linkage of the left engine throttle mechanism. The stainless-steel oil pan had been reduced to a molten mess by the extreme heat.

Any further delay to our landing would have had severe consequences.

Ignorance is bliss, but I handled that precarious emergency in Ice Man-like fashion.

That was the old me. I'm starting to miss him.

*

Our flight proceeds without further incident. 'Welcome back, Kev,' my training captain announces after a smooth landing and safe arrival at our parking gate. He compliments me on my aircraft management and manipulation.

I'm quiet and perplexed; my triumphant return to my passion feels like walking into a party with all the balloons deflated. *Give it some time*, I coax myself, trying to cut myself some slack. After all, it was my first trip back after eight months off, in an aircraft that had behaved maliciously without all the causes fully identified.

But I'm concerned by my twinge of body response in the turbulence. Before QF72, nothing in my commercial aviation career ever phased me – now the sensation of negative g caused by the autopilot has triggered a reflex previously foreign to me. Because it didn't negatively affect my performance, I put it in a mental compartment for later inspection.

Before I go to sleep, I quickly message my daughter to let her know I've survived my first supervised flight without incident.

*

On the bus from our hotel to the night flight back to Sydney, my training captain passes me a sheet of paper. 'What do you think?'

It's the pre-departure summary that notes any special loading, passenger requirements, delays and maintenance issues. It also designates the registration of the aircraft we will fly. As he watches me intently, I read the rego letters of our plane: VH-QPA.

'So?' I reply. 'QPA is probably the best A330 we have now. All the computers have been stripped and replaced with new ones. Are you worried?' I shrug. 'I lost my uniform tie on this plane, so maybe I can find it now.'

I have an objective approach to life and to stress – I don't generate drama or emotion and will brutally strip it away when making big decisions.

Unfortunately, one of the cabin crew overhears our short discussion. 'What's wrong with this aeroplane?' he asks, sounding superstitious.

With a hand gesture I defer the question to the training captain; he initiated the drama, so he can defuse it.

'Oh, this was the aircraft involved with QF72,' he says. 'It's okay now.'

'QF72? What was that?'

I raise an eyebrow at my training captain and leave it to him.

It's a scene that will be repeated many times in the years ahead.

*

After two and a half hours of uneventful cruising, it's time for my break. I brief the training captain on our progress, fuel usage, weather and closest divert airports, then I exit the flight deck. My rest routine involves a walk through the cabin, visiting with the crew and checking on the passengers. I stop in one of the galleys for a drink of water and a chat before navigating through the darkened cabin to our pilot rest area.

With frustration, I notice that most passengers have unclipped their seatbelts while they sleep, despite my standard public address and the pre-departure safety briefing.

The pilot rest facility has no windows so it gets very dark, with only one or two small indicator lights burning faintly. The bed isn't that comfortable after years of use, but at least we can lie flat. I listen to the dull drone of the engines before inserting some earplugs, fastening my seatbelt and drifting off.

*Ding.*

My deep sleep is disrupted and I wake with a jolt. I hear the characteristic master caution *ding* on the flight deck all the way through the bulkhead wall of the crew rest and my heavy foam earplugs. I sit upright in the bunk, remove the plugs and strain to hear any other warnings or chimes.

Then I pick up the interphone.

'Flight deck,' comes the answer to my call.

'Hey, it's Kev. Everything okay up there?'

The two pilots chuckle.

'Did you hear that?' the training captain asks, sounding surprised.

'Yeah. I'm pretty sensitive to that sound for obvious reasons. Is everything okay?'

'Relax, Kev, it was a minor cabin temperature issue and we've sorted it out. Go back to sleep!'

I hang up and lie back in the bunk. *How did I hear that?* I've never done that before and rarely micromanage operations on the flight deck from the crew rest. Another reflex change in my psyche has identified itself, raising my level of confusion. *Why is this happening?* My first trip back is evolving into a journey of discovery – of the new me. Perplexed, I try to go back to sleep.

As the rising sun glows below the horizon, I return to the flight deck. We've passed New Guinea and soon will be talking with Brisbane Air Traffic Control as we continue south on our track to Sydney. The air mass is still, we're in clear air and the pre-dawn lightshow is on display. The horizon is a watercolour of vivid hues contrasting against the blackness of the sky above, the glowing reds, oranges and yellows of the sun's rays projecting into the atmosphere.

I crane my neck around and look back towards the wingtip at the indigo darkness of space to the north, sparkling with the stars and planets, illuminated by the light of the approaching dawn. I've missed this time of day during my extended break from flying; it's a special treat to view the changing light from 39,000 feet. The sun is rising rapidly, the sky changing quickly and brightly, prompting me to put on my sunglasses. I know I still belong here.

Our approach into Sydney is smooth and uneventful. QPA has behaved itself, and my training captain has very few comments on my performance as we exit the aircraft. My passengers and crew are happy and relaxed as I observe them dissolve into the arrivals terminal. But I'm feeling hyped and fatigued. Despite the routine arrival and landing, my body hasn't responded appropriately. The nice buzz I used to feel at the completion of a flight has been replaced with something I can't describe. It's similar to the feeling I had when we landed in Learmonth; it's the fight-or-flight body response rather than the glow of an endorphin release.

*

After a few days off, I'm scheduled for my final supervised flight to cement my return to line operations. It's very routine: fly to Perth, one night off, fly back. I fly as well as I ever have and pass with minimal comment.

In August 2009, I'm cleared back as an operating captain. Although I'm not elated about being requalified, I'm certainly relieved – I've made a herculean effort to return to the A330, in the face of an ongoing investigation into a failure of titanic proportions. Things *have* changed, though. Can you ever comfortably walk down a street where you were attacked? Can you ever sleep soundly in any house where you've survived a violent home invasion? Faced with a near-death or violent event that makes you question whether the outcome will be life or death, is life ever the same?

Now that I'm back in the air, in accordance with the self-reporting policy it's up to me to raise a flag and identify myself if I am experiencing any health issues. If you don't identify yourself, then everything is okay … right? I'm an accidental hero, left alone to travel a hard and unchartered road to recovery. I must use what I have to survive. It reminds me of the way I had to use the systems that were still working to land in Learmonth. I have my health, my job and my sense of humour. My daughter still loves her dad, and my core group of friends are still supportive. I haven't experienced the darker, more extreme behaviours that PTSD can produce. This is the choice: to be a survivor and not a victim, and to rebuild.

My colleagues are still peppering me with questions on the particulars of QF72, when all I want is to re-establish myself in the air. It all starts to whittle away at me, and I find myself feeling exhausted at the end of every flight. It's like getting in a boxing ring with Iron Mike Tyson – dancing around the ring and trying to dodge his punches is slowly wearing me down.

On 18 November 2009, the Australian Transport Safety Bureau publishes a media release about the ongoing investigation.

It brings up two significant points: an update to the computer software by Airbus, and an opinion regarding the AF447 accident investigation.

The Airbus software updates address the algorithms that caused the dives. They also improve the way air data is processed by the flight-control computers from the air-data computers. The installation period will occur over the next twelve months.

The ATSB compares AF447 and QF72, addressing concerns that the two are related. The conclusion is that there are no similarities, because of a difference in the ADIRU models and the sequence of failures that generated a problem with the aircraft's displayed airspeed. I don't agree – I believe the weak link in the Airbus design is being revealed. It reminds me of what happened to the XL Airways flight. If the sensors become blocked or damaged, they'll send erroneous data values into the ADIRU; the ADIRU will package this value and distribute it as it's designed to do.

I look at it simply: if a dog turd goes in to the ADIRU, it's processed then packaged, and a dog turd comes out in a digital value from the ADIRU. This is then distributed to all the user systems, generating a tsunami of poo – in this case failures. Erroneous speed or altitude is the by-product and difficulties can escalate if the computers activate protections. The pilot is left confused and overloaded in a life-or-death battle to save the day.

This is on my mind each time I step back into the A330.

# 26.

Public expectations of pilots are very high. What other occupations command the same levels of responsibility and performance, or have such a direct influence on the safety of several hundred people at one time? It takes a massive leap of faith to purchase an airline ticket and confine yourself in an aluminium tube for hours, traversing through an extreme environment with turbulence and thunderstorms, to be safely delivered at your destination. Routinely, pilots are ranked as having one of the top trusted occupations in the world, along with paramedics, nurses, doctors, law enforcement officers and firefighters.

Modern society places enormous pressures on all of us. Most people want a home and a family, to enjoy a loving and secure relationship, and to excel in our chosen professions. For a lot of Western men, 'failure to achieve' is the greatest fear driving us forward. Many of us have been groomed from childhood to perform, to suffer in silence and to not show our emotions. All of this sets us up for problems when our goals come unstuck, paving the way for depression, medical complications or all of the above.

I've watched many interviews with combat veterans from

both sides of conflict in countless wars, who are now in their senior years. They always react emotionally when asked to recall what they had to do in war and the comrades they lost – fifty, sixty, seventy years after their ordeals. They had their innocence torn out at an early age, and they've been forced to manage that for the rest of their lives.

These warriors are from a generation that required traumatic memories to be suppressed. I witnessed this in my father, but at the time, I didn't really comprehend the sensitivities involved. I have already decided that I will not walk in these footsteps.

Pilots are expected, from early in our careers, to self-analyse and self-critique as we perform our technical duties. But this process may fall short when it comes time for us to check our emotional pulse and to accept that our stresses cause distractions that may, at some stage, adversely affect our ability to perform.

As a military pilot, I soon learned the fine art of compartmentalisation. We had to be clinically brutal in mastering and employing this mental process to survive our dangerous environment. The inability of an officer to manage stress doesn't bode well for professional progression. Slamming an unforeseen complication into a mental box is a skill set that ensures survival in the short term – but at some stage, the gooey contents must be removed from that box and managed appropriately.

Building up a collection of leaky mental boxes without emptying them routinely paves the way for serious social and psychological issues.

I'm a master compartmentaliser, and it's time I open up my QF72 boxes.

In early March 2010, I twice have to deal with unusually severe turbulence while the autopilot generates significant negative g to maintain altitude. I manage these events effectively and safely, but my body doesn't. I'm holding on to the glareshield again as my copilot lounges in his seat and observes my body language with a perplexed look.

On 8 March, I self-report to management. I can't continue to fly responsibly in light of the distractions I'm dealing with. To examine the contents of my leaky boxes, I need time away. I hate being grounded again, but I have to try to recognise the triggers of these body responses, and to come up with strategies to manage them. I could have tried to endure my hell in silence, but I know it's best for me to be open.

I look upon my decision to stop as courageous and responsible. A pilot doesn't have a leg to stand on if he's hiding psychological issues and mismanages an operational event in the course of managing his own stresses.

I don't know how much time I'll need to figure out what's causing my body responses. I know I'll need professional advice, so I consult two psychologists, who assure me that this post-trauma condition should diminish in time. But, as the weeks go by, I feel more burdened. I feel like it's up to me to sort myself out and raise any concerns.

Following sustained self-examination, particularly based on hours in the cockpit, I finally work out that my body's response to negative g has been rewired.

The autopilot's operation in turbulence can be abrupt; whenever the pilot commands an altitude to be maintained, the autopilot will carry out that command aggressively with a pitch correction. The magnitude of the vertical displacement is met with an appropriate correction, and negative g can often be generated as the autopilot pushes the aircraft back down to the commanded altitude.

My body is automatically reacting to the negative g in turbulence as it did during the computer pitch-downs. To my parasympathetic nervous system, which is outside my conscious control, negative g equals loss-of-control and mass casualties. Exposure to negative g, in any form, generates a startle response that instantly sends a mini-dose of fight-or-flight chemicals through my body. The same is true for my body's reactions to the various warning and caution alerts.

All the ones and zeroes of my emotional operating system have been modified, and not just in a professional context. For the first time in my life, in many social situations I'm experiencing hot flushes and a beet-red face, and the feeling is like being embarrassed on steroids. And, unfortunately, these emotional episodes are becoming the norm.

Absorbing the stresses and challenges as the captain of QF72 was akin to jumping on a hand grenade. I saved my passengers and crew, survived the explosion but sustained critical damage to my psyche. A big hole has been blown out of me and it's not going to fill itself in. Perhaps the best healing I can hope for is scar tissue lining the rim.

It's a small relief that I've identified the triggers. They aren't dissipating, and I'll have to face them on practically every commercial flight.

*

Qantas gets in touch with me, Ross and Pete. They'd like to recognise our handling of the QF72 accident with an award.

The Excel Award routinely acknowledges excellent performance from staff in the line of duty. We accept the offer and attend an awards ceremony celebrating those who have received an Excel.

Ross and Pete bring their wives, Mel and Claire, to the ceremony. Mel and Claire have endured eighteen months witnessing the event's impact on their husbands.

After we've accepted our award and return to our table Mel and Claire give us a standing ovation. These loyal women want to help their hero husbands, but don't know how. Still, later that evening, they speak to two managers and tell them they want their husbands back to what they were before the event. It reminds me just how destructive PTSD can be, not only for the sufferer but for their loved ones and colleagues too.

There is a more prestigious award – the Chairman's Award. My crew deserve to receive it, and in 2009 I try to broker it for them. They would have worn their pins proudly on their uniforms every time they went to work (for those who were able to return) and received the accolades from their fellow cabin crew that they deserve.

But if you aren't wearing a Chairman's Award pin, no one can ask you what happened, I guess.

I've never worn my Excel pin. It lurks somewhere in a corner of my sock drawer.

# 27.

I'm cleared to fly again in August 2010. My professional performance remains above average, despite the gradual decline in my personal life. But the slow rot of PTSD is manifesting itself. If I can't develop different strategies to manage my body responses, the demise of my career is on the table.

In May 2011, the Australian Transport Safety Bureau contacts me. The draft of their final report is available for review and comment from the interested parties. I'm desperate for the investigation to be finalised so I can move forward. I'm not worried about the results but I want closure. And I want to know that all the causes have been identified to renew my trust in the design. The previous reports haven't commented on my decisions or the performance of the crew and, naturally, I'm anticipating their findings on these topics.

I review my draft of the ATSB final report and am impressed with its scope and objectivity. Finally, I'll learn exactly what went wrong and what has been done to prevent it from recurring.

The ATSB interim and preliminary reports identify and summarise the root causes of the computer failures. Firstly, air-data values from the ADIRU 1 had somehow been corrupted during the processing phase and distributed to the primary flight-control computers (PRIMs). Secondly, faulty algorithms installed in the computer software permitted the corrupt data to be used as valid; this made the PRIMs activate protection modes in response to the erroneous data. Up to this point, the workings of the ADIRU and the faulty algorithms have not been discussed but are now explained in the final report.

Their report is 291 pages long, and the language is very technical, in keeping with the extremely technical failures of the QF72 accident. These are powerful computers, processing data and controlling all the functions and systems on the aircraft. The findings are based on three years of analysis of the recorded data from the black boxes and extensive coordination with the two manufacturers involved in the investigation, Northrop Grumman and Airbus.

The findings are a revelation to me because the systems' complex behaviour and the commanded manoeuvres by the flight-control computers are mostly transparent to the pilot. The barrage of failures that occurred over a seven-and-a-half-minute period are summarised in a report that took three years to complete.

*

I arrange a face-to-face discussion with an ATSB investigator, Sam, to review the report and to finally listen to the cockpit voice recorder. Over the past three years, the ATSB team have been extremely professional and cordial, always available to answer my questions and hear my concerns.

In the preamble of the draft report, the role of the ATSB in any investigation is described:

> It's not a function of the ATSB to apportion blame or determine liability. At the same time, an investigation report must include factual material of sufficient weight to support the analysis and findings. At all times the ATSB endeavours to balance the use of material that could imply adverse comment with the need to properly explain what happened, and why, in a fair and unbiased manner.

This is the starting point of my discussion with Sam. The ATSB aren't there to blame, but to analyse and report objectively in the interests of safety.

Sam is an intelligent, softly spoken man with enormous patience. As he methodically steps me through the report and explains the complexities of the failures, my initial assessment of the extreme nature of the accident is validated: this highly automated, complex machine generated complex failures that were unforeseen and untested during its development.

Sam and I match the timeline of failures with the timeline of the flight. Remember that I returned to duty at time 1239 in Perth. What caused the first pitch-down?

The first pitch-down occurred two minutes after I returned to the flight deck. A lot happened in those two minutes – the shit started to hit the fan as soon as Ross returned from the toilet.

Hidden from the pilots, corrupted data was being randomly generated from ADIRU 1 in both the air-data and inertial reference partitions of this computer processor. The report labels them as 'data spikes'. These caused the autopilot to disengage.

The first of many aircraft systems started to register faults, and the failures started to appear on our centre displays. The corrupt data then satisfied the criterion to generate stall and overspeed warnings. My airspeed and altitude displays were also affected, making them jump up and down on my primary display. This caused confusion but we addressed some of the faults while I was forced into manual flight due to the unreliable speed situation.

In the two minutes before the first pitch-down, twenty-two master caution chimes, seven overspeed warnings and ten stall warnings were recorded.

I wasn't counting them but the aircraft's recorders were. You can do the maths: thirty-nine warnings and cautions in two minutes, with stall and overspeed warnings indicating a serious flight condition that requires pilot action to remedy. Are you *kidding*?! As mentioned previously, these warnings and cautions couldn't be silenced, which generated a tsunami of confusion and sensory overload.

During these two minutes, I was 'sitting on my hands' and trying to analyse what was happening. I knew I wasn't stalling or overspeeding because, seconds before, we'd been in a stabilised flight condition, and nothing had changed to upset that stability. I could also confirm that we weren't at the stall angle of attack, despite the incessant stall warnings, because I could look outside and visually verify that our flight path was safe. I wasn't overly concerned at this time because the Airbus technical manuals don't mention that erroneous stall warnings can generate a violent computer response. Perhaps they should.

Sam and I discuss my concerns and actions with the unreliable speed problem. There was a rapid, memory-based procedure for this condition, and the first step was to disengage the autopilot and manually fly. I didn't want to remove the assistance of the autopilot while I was trying to find the causes of our problems, but it was a requirement. The first pitch-down would have occurred whether the autopilot was engaged or not.

At the speed of light, myriad data transmissions, computations and computer commands were flowing through the wiring between the air-data computers, the flight-control computers (PRIMs) and the aircraft's elevator control system. The flight-control computers were looking closely at the bad angle-of-attack data spikes being produced in ADIRU 1.

Because it was extreme, the PRIM activated a 'testing' algorithm within its software, to test that value for authenticity. The extreme rate of false data spikes streaming from ADIRU 1 satisfied this algorithm. The erroneous value of angle of attack passed the test and was confirmed as 'valid' by the flight-control computer.

Then the PRIM activated two modes to protect the aircraft from a false conclusion that it was about to stall. A nose-down command was transmitted to the aircraft's elevator in the tail. It forced the nose down in an abrupt, accelerating, pitching motion, then abruptly it commanded the nose up.

Classified as a pitch-down, it was like a boxer's one-two punch or one of those novelty drinking birds, but on steroids.

The computer's pitch commands were rapid and violent, generating destructive g-forces and altitude loss.

My sidestick was blocked and ineffective for 1.8 seconds. The pitch-down generated a loss of 150 feet in that time.

In total, I used 690 feet of altitude to recover and level off before climbing back to 37,000 feet.

One of the flight-control computers had faulted during the first pitch-down, PRIM 3. The stall warnings were silent during this faulted period until it was reset.

The data spike production from ADIRU 1 continued, the warnings recommenced and our confusion was amplified.

The second pitch-down wasn't as violent as the first, but it still generated an altitude loss and a lifting g-force.

The investigation provides evidence that the PRIM algorithm again tested the data spikes and confirmed the value as valid. The computers sensed a threat to the aircraft, and the data spike behaviour forced a change in the level of protection available from the flight-control system.

As a result, only one protection was activated.

The aircraft's data shows my sidestick was blocked for 2.8 seconds until the completion of the protection manoeuvre. I regained control in my sidestick and started climbing back to 37,000 feet. We lost 400 feet during the dive and recovery.

The flight-control computers had reconfigured themselves automatically during the pitch-downs after fault conditions were generated. PRIM 3 faulted again and was displayed to the pilots. The level of aircraft protection was changed to Alternate Law because of this transparent reconfiguration, but it wasn't displayed to us at the time. No information was displayed to notify the pilots of protection mode activation, data spiking or the specific protections activated. (In the report, the explanation describing this reconfiguration is complex.)

The confusing stall and overspeed warnings continued.

I notice that the automatic pitch trim wasn't working after the second pitch-down, and that I had to operate it manually. Sam explains that the loss of the automatic pitch trim was due to the reconfiguration of the flight-control computers. Its failure wasn't communicated to the pilots.

The master caution chime was activating continuously in response to multiple system failures and faults. The fault messages were displayed momentarily and then disappeared, each replaced by a new fault. We couldn't action any of them. Sam explains this was due to the rejection of air data from ADIRU 1 by many of the aircraft's automated systems. I comment that this was an enormous addition to our workload.

QF72 then landed safely at 0532.

'How exactly were the data spikes produced?' I ask Sam.

He explains the technical failures in the Airbus software (Section 2 of the report) and the Northrop Grumman air-data computer (Section 3). The discussion is very technical, and I won't attempt to describe it. The most important thing to note here is that the investigation reached two conclusions: without one failure, the other would not have occurred. The air-data computer generated extreme data and sent it to the flight-control computers. The PRIMs tested it then used it to injure 119 people on QF72. This is a simple summary for a tangled mess of failures.

Airbus has since been able to identify these particular flaws in their software and modify their algorithms so that this specific failure scenario won't happen again. The investigation couldn't pinpoint the reason why the Northrop Grumman air-data unit took good data, corrupted it and hid its faulty operation from the pilots. That remains unanswered.

One thing is certain: if the faulty unit had identified itself and been turned off, the accident would have been averted. But there was no procedure in place to address this unforeseen failure sequence in the ADIRU and its effect on the flight computer's software, because it had not been fully tested during its development. There is a procedure now, but it isn't an intuitive action like pushing a button on the control stick or forward instrument panel.

In 2010 Airbus ended their partnership with Northrop Grumman, and switched to the Honeywell corporation for the supply of air-data computers. Likewise, starting in 2011, Qantas replaced all their Northrop Grumman air-data computers with those from Honeywell. It took a couple of years to complete the refit.

We survived a perfect storm of automation failure. I've already concluded there was nothing more I could have done to stop the injuries dished out by my confused Airbus A330, and this debrief with Sam has confirmed my assessment – but I don't find comfort in that realisation. This accident damaged me and exposed me to the unforgettable images of my injured passengers and crew. The closure I've desperately sought is denied.

The discussion of the ADIRU operation and the complexity of the PRIM algorithms does not renew my confidence in the

Airbus design. The achievement of the pilots working together to save QF72, while noted, is not redeeming. I already knew we had come back from the brink.

*

It's time for me to listen to the cockpit voice recorder. It contains all the conversations, sounds and warnings in the cockpit and from our radio transmissions, and also from inside the cabin through the intercom system. The recording is a protected document in Australian aviation investigation and legislation, and I honour that element of privacy; in many countries, it isn't protected.

As I sit there with Sam, I recall the emotional effect the recording had on Pete and Ross when they first listened to it. Sam warns me of its confronting content before he pushes 'play'.

My recollection of the flight is different to what I'm hearing from the recorder. I've told many friends that I was calling this Airbus every bad word in the book as its malfunctions threatened everyone on board. But I don't hear any of this from the recorder; in fact, if you were a casual observer unaware of what was happening in the cockpit or with the aeroplane, you'd be excused for thinking that Ross and I were sitting on a park bench, quietly feeding the pigeons. We sound calm, professional and focused, despite the massive doses of adrenalin coursing through our bodies. Ross's 'What the fuck was that?!' is recorded truthfully and accurately, however.

The cacophony of warnings and alerts is captured. Each one required recognition and reaction, but which ones were real and required action, and which ones weren't?

The most chilling sound is the *crash* I heard during the first dive. It sounds like a 40-foot shipping container full of plates and glasses smashing into a brick wall. I can clearly hear the passengers screaming as they were catapulted into the ceiling. I can hear the forward galley's meal carts and storage boxes exploding onto the floor as the plane lurches down. The screams stop abruptly; the passengers had hit the ceiling, stunning them, before they were thrown onto the floor. Their screams are replaced by moans and sobs. It's a morbid recording.

Sam is watching me closely. I take a deep breath and exhale. It was hell in the cabin, a terrifying hell. It was an ambush from an old western movie, and the settlers were being scalped alive.

But, despite the confronting nature of these sounds of pain and despair, I need to hear them.

'Yeah, Sam. It was hell,' is all I can muster in response to his inquisitive gaze.

He fast-forwards to a particular point that he has marked for my listening: Pete was talking on the satphone, Ross was on the intercom talking to the cabin crew and reorganising our flight libraries, and I was manually flying while talking to Melbourne Air Traffic Control. The ever-present dings and warnings are recorded too.

Sam stops the tape. 'Your workload here is enormous,' he remarks.

'Yes, it was. We had to separate to complete these independent tasks but everyone still had to crosscheck everything I was doing while I was manual flying, and to try and keep track of the information being received from the maintenance unit and the cabin crew. All those tasks were required to aid our preparation for approach and landing.'

Unfortunately, this workload can't be replicated in the flight simulator; if you can handle this scenario then you can handle anything.

Of course, I remembered almost everything that we did and said, which is recalled in previous chapters based on my memories on the day of the accident. The rest of the tape leaves me disappointed – I still can't hear any of my foul-mouthed derision and scolding of the aircraft and its bad behaviour, but I certainly didn't hold back.

My meeting with Sam is over, and I thank him and his team. Their report should be documented as an industry standard for its complexity, objectivity and attention to detail.

*

When I read the final ATSB report, I'm relieved to see this comment: 'The flight crew's responses to the warnings and cautions, the pitch-down events, and the consequences of the pitch-down events, demonstrated sound judgement and a professional approach.' I didn't discuss its inclusion with Sam.

There's much more in this section. It justifies our decision to divert to Learmonth and our distrust in the failing operation of the many computer systems. There's no criticism of the pilots, and I'm pleased to see a thumbs-up for the heroic actions of the cabin crew.

Case closed. Not for me. The images of the carnage and injuries remain imbedded in my memory.

Section 4 of the final report analyses the forces generated and summarises the injuries. This is the first time I'm aware of the serious nature and scope of the injuries experienced in the cabin.

It mentions that two passengers were seriously injured, even though their seatbelts were fastened: one was a child admitted to hospital with abdominal contusions; the other was an adult who experienced neck pains and was admitted to hospital three days later after having a stroke.

There was damage in some of the toilets, and I remember noticing this after we landed at Learmonth. One was completely unhinged during the pitch-downs while someone was using it. Imagine the injuries that this person must have endured after being slammed into the ceiling by the negative g, then being thrown to the floor with the dislodged toilet assembly under or on top of them.

Hidden at the bottom of one page is Footnote 199. It describes the serious injuries inflicted upon Fuzzy during the pitch-downs: 'The flight attendant who was standing in the rear cabin impacted a handrail in the galley after hitting the ceiling, which exacerbated his injuries and also damaged the handrail.' It's an understatement to say, 'and also damaged the handrail'. His body struck the stainless steel handrail with such force that it deformed the railing; this impact made him return to consciousness for an instant before his knees were slammed into the galley floor.

Due to the uncertainty of further manoeuvres occurring, I was forced to keep everyone in their seats with seatbelts fastened. The cabin crew instructed passengers to clear the aisles of debris to prepare for the likelihood of an emergency evacuation once we landed. This is incredible initiative:

> Prior to and after the second event, the cabin crew provided instructions to passengers to be seated and to keep their seat belts fastened. Some passengers requested medical assistance but the cabin crew advised them that they were unable to leave their seats. Some passengers provided medical attention to other passengers seated close to them, and cabin crew provided advice from their seats to some passengers about medical treatment. Towards the end of the flight, cabin crew and passengers in some areas cleared the aisles of debris and bags within their reach and without leaving their seats.

The report summarises the physical injuries sustained by the people in the cabin, but it doesn't mention the enormous psychological stress they experienced for the fifty minutes we were airborne and preparing to land after the first dive.

Our situation makes me think of movie scenes where a soldier or a kidnap victim is kept in a high state of uncertainty as their captors perform mock executions with unloaded weapons held to their hooded heads. I feel this accurately represents the stresses experienced by everyone onboard during those fifty minutes.

I walked out, thankful for retaining my scalp, but I'm still burdened. We had so many factors working against us, and again I felt lucky to have survived.

I consider the QF72 accident to be one of the most catastrophic in modern aviation. All the automated systems, designed and installed for enhanced safety, failed. A complex system created a complex and destructive failure.

# 28.

We're in the age of automation addiction.

There's no doubt that automation in aviation has saved lives. Advances in software design and performance, fly-by-wire and electronic flight-control systems, and flight simulators have also generated savings to airline operators competing for profits and market share.

But I'm a survivor of an automation disaster, and I have a responsibility to issue a warning.

If increased automation is incorporated into an aircraft's design, then it must be a good crew member. It should enhance and improve the human pilots' operational effectiveness and aircraft management.

Unfortunately, on QF72 we had a bad crew member. It presented us with incorrect information and distracted us with erroneous warnings that couldn't be silenced. It refused to cooperate and denied interaction. After the second pitch-down, its faulty systems identified themselves until they overloaded the aircraft's ability to display and refresh this distracting flood of information – but it didn't display the faulted operation of

ADIRU 1. Its PRIMs commanded the simultaneous activation of two protections that had a combined input to the aircraft's elevator, something that is omitted from the manufacturer's manuals; how many protection manoeuvres can the computers add together after sensing a deviation from normal operation?

A pilot's startle response can be provoked by an automated system that doesn't warn of the impending activation of protections – protections that can generate violent manoeuvres in its confused state and prevent pilots from intervening with intuitive control inputs. Such design flaws can increase workload and cause bewilderment for the crew responsible for managing the aircraft.

Yes, I label them 'flaws', not 'limitations'. If the automation is to be an enhancement then its design should allow the operator to troubleshoot problems with speed and confidence. In the case of QF72, the design generated a complete loss of confidence in the automated systems.

Automation is a sleeping pill for today's pilots. It hypnotises them into thinking, *I can relax because everything is being monitored, and the automation will tell me what I need to do.* Pilots now wait for automation to wake them up, and to reconfigure systems in response to faults or failures. Many airline operators perpetuate this erosion of skills by requiring their pilots to fly exclusively with the automation active.

In 2013 Asiana Flight 214, a Boeing 777, impacted the seawall on approach to San Francisco Airport. In this situation, automation was desired but couldn't be used for the approach flown.

The ground-based instrument landing system that links to the autopilot was out of service, forcing the pilots to fly a visual, manual approach with only the automated engine thrust available. A lack of understanding in the operation of the auto thrust by the pilots generated a low-energy condition prior to landing. A contributing factor was that, quoting the final NTSB report, 'Asiana's automation policy emphasised the full use of all automation and did not encourage manual flight during line operations.'

This new generation of automated aircraft have an Achilles heel. The suite of external sensors and probes – which provide the critical values of speed, altitude, temperature and body angle – are vulnerable to the extremes of the air mass in which they operate. The generation of false unreliable speed by blocked sensors is potentially a dynamic and confusing scenario on an Airbus. The associated software may not be capable of recognising its veracity. Human pilots are required to recognise unreliable speed through their interpretation of the instruments. If this assessment is inaccurate, or the pilot is distracted by simultaneous failures or conditions such as turbulence, then aircraft management can degrade rapidly.

When unreliable speed is practised in the flight simulator, it usually involves a singular, uncomplicated scenario that causes the pilot's primary display to 'blank'. This isn't simulating an environmental degradation of speed, only an electronic failure of the display. That's completely unrealistic and doesn't prepare the pilot for the cascading problems caused by corrupted data.

In this non-normal scenario, pilots are directed to disconnect the automation – autopilot and autothrust – which they rely upon to make their job easier. Manual flying is the forced result. Pilots don't practise high-altitude manual flight in the aircraft, and manual flight isn't an easy task in Alternate or Direct modes of operation at high altitude. Throw in night, bad weather, convective activity or turbulence, and the workload increases markedly. The reconfiguration procedures to address these systems failures, although automated, are time-consuming and distracting, and the management and control of the aircraft can quickly be compromised.

An example of this sensor failure occurred on Air France 447. While the aircraft was flying through thunderstorms, its pitot probes, which measure airspeed, were blocked with ice crystals. This also generated a false altitude loss display to the pilots. The corrupted air data, produced by the blocked probes, forced a disconnection of the autopilot and autothrust. The pilot, through startle response and in reaction to the falsely displayed altitude loss, pulled back on his sidestick. The aircraft climbed rapidly and exceeded the normal pitch parameters, resulting in the automatic removal of protections. The pilot was now permitted to stall the aircraft. The supposedly stall-proof Airbus was stalled so badly that it was described as a 'super stall'. The flight crew didn't have the training or experience to recognise their situation and recover from this extreme state.

This accident was caused by a combination of automation failure and inappropriate pilot response. It could have been avoided.

Information from the flight data recorder and cockpit voice recorder was publicly provided by the state investigative body, the

French BEA. Reading the voice-recorder transcript sends chills down my spine, strongly reminding me of QF72.

Embedded in the BEA report are some significant findings. Regarding unreliable speed events and training, the report notes: 'the number and the type of manifestation linked to erroneous speed indications makes training and exhaustive information to pilots impossible'. The report also comments on the link between displayed error messages and the pilots' ability to determine the unreliable speed situation: 'In the absence of a specific message expressing detection of unreliable speed by the systems, the crew was unable to identify any logical link between the symptoms perceived and these ECAM (fault) messages.'

Basically, the report accepts that a void exists in the aviation industry regarding training for unreliable speed scenarios. However, the report doesn't criticise the Airbus design and its contribution to the fatal outcome. The BEA is a government agency, and participating governments from the European Union subsidise the Airbus consortium.

There are striking similarities between the QF72 and AF447 accidents: unreliable speed, pilot overload, flight-control reversion to Alternate Law, automation-induced confusion, increased workload and the pilots' startle response. How the two crews got there is quite different, but the destination is the same. If the Air France crew had 'sat on their hands' and not reacted as aggressively to the unreliable speed indication that was recorded, then it would have been a non-event.

I won't be an armchair quarterback and comment adversely on these pilots' actions, their response to an abnormal event they hadn't experienced before. It's a tragedy, and they fought gallantly to regain control once the captain returned to the flight deck, but they ran out of altitude.

NASA also conducted an analysis of the AF447 accident, concluding that the automated systems generated more abnormal failures than the pilot is trained to cope with.

*

Automation comes with side effects that are often unintended, and accidents continue to happen. Many of them are now caused by confusion in the interface between the pilot and an automated machine programmed by a human.

A goat in a tuxedo is still a goat. In the words of Earl Wiener, an engineer who taught at the University of Miami:

- Every device creates its own opportunity for human error.
- Exotic devices create exotic problems.
- Digital devices tune out small errors while creating opportunities for large errors.
- Invention is the mother of necessity.
- Some problems have no solution.
- It takes an aeroplane to bring out the worst in a pilot.
- Whenever you solve a problem, you usually create one. You can only hope that the one you created is less critical than the one you eliminated.

To ensure pilots understand all the complex scenarios that can be generated through displayed failures, their training has to extend well beyond normal modes of operation. If not, confusion can develop, causing pilots to question – most times in seconds – what's real and what's false, which diminishes their ability to make necessary interventions. It should be the manufacturers' responsibility to provide enough training for pilots to be confident in all the system's modes of operation. If the manufacturers don't define which scenarios present an unacceptable risk, what chance does a human pilot have to intervene effectively when the automation fails?

Airbus states that the piloting experience required to operate their equipment can be minimised due to the enhanced safety afforded by increased automation and ease of manipulation. But the Airbus design has a history of provoking confusion and startle responses, so is that really the place for an inexperienced pilot? One thing's for sure: a novice can easily be taught basic manipulation and operation in only a few simulator sessions. But aviation, even in a protected fly-by-wire aircraft, is unforgiving of error. How will an airline operator manage a pilot's degradation of core manipulation skills if the pilot begins their career with a minimum experience base?

On QF72, my fallback position of extreme military flying experience was pivotal in our survival. So, too, for the other Sully.

Is there a need to train flight crews for loss of automation? The answer is yes, now more than ever. It's also time for manufacturers to share information with simulator developers so they can create more useful modules for pilots.

Fly-by-wire leads to the erosion of basic flying skills, along with a remarkable increase in pilot overconfidence. This isn't my unique opinion, but a view expressed by regulatory organisations and aviation professionals alike. Automation bias and a reduced motivation to explore issues outside the normal grind of commercial flying make it easier for a pilot to think, *I'll deal with it on the day.* Automation events can then push pilots well out of their limited experiences and abilities. This is part and parcel of the new age of automation reliance.

Maybe their egos are writing cheques that their abilities can't cash.

Pilots flying the new generation of electronic flight control and envelope-protected aircraft are conditioned to manage flight-path tasks with a pristine, by-the-book mentality that doesn't give much consideration to unexpected failures of the installed technology – even though, historically, there are plenty to choose from, most occurring in an Airbus. It's also easy to dismiss concerns with the causes of automation events after the manufacturer has said, 'We fixed that.'

If you're a pilot, especially one operating on this particular piece of equipment, and you think you don't have to worry about it – well, I have some sobering news for you. You should worry about it, and so should your training departments. Simply relying on 'book knowledge' provides a false sense of security. The books can't teach the decision-making and time-management skills required to manage the overload of simultaneous failures in automated systems. The books don't delve into the software's complexities and the automated systems' reliance on air data, as these topics aren't considered need-to-know information for the pilots who fly these aircraft.

Software should be enhanced to help pilots identify unreliable speed, and it should effectively deliver critical fault information through its warning systems and displays. Warnings should be designed to minimise pilots' startle responses and increase their situational awareness. Angle-of-attack data, a critical value to fly-by-wire aircraft, should be displayed to the pilots as standard equipment.

Pilots must be trained to manage startle. They should also be taught to manage the overload of multiple, complex system failures as recorded in the QF72 and AF447 accidents. And they should train for a loss of automation to ensure their core skills are present if manual manipulation is needed to recover a degraded aircraft. In 2019, the US Federal Aviation Administration mandated this training for all commercial aircraft operators. Australia will follow suit.

Meanwhile, manufacturers are driven by traditional statistics from the accident history of previous-generation aircraft to design systems with increased automation while minimising direct pilot manipulation. After all, it's a question of securing market share in providing global infrastructure, and in this instance, it goes beyond Airbus versus Boeing to the degree of the EU versus the US.

Are manufacturers and airline operators loathe to admit to serious issues with their machines because of liability? If so, are passengers and crew being regulated to the bottom end of the corporate food chain, with acceptable losses factored in by risk analysis and actuaries? Perhaps it's just plain corporate greed, and we're expendable in order to protect brands and profits.

For most of the parties involved, it's very convenient to praise the automation and blame the pilot. But as a survivor, I know on which side of the fence my opinion lies. If you believe the marketing departments of automated systems, then you must accept that humans are unsafe operators of aircraft – and of motor vehicles.

*

I read an estimate that the autonomous vehicle industry is worth US$1 trillion – a juicy carrot for the many technology companies racing to get their products to market.

Bolstered by an increase in motor vehicle-related deaths, the 'driving' force towards automation in cars is the goal of reducing accidents and enhancing safety. In the US, it's estimated that 41,000 people lost their lives in motor vehicle accidents in 2017; the year before, it was 37,000. In China, the World Health Organization estimated that 260,000 deaths – including pedestrians – were attributed to road-traffic accidents in 2016. Worldwide, that estimate is 1.25 million deaths due to road-traffic injuries. It's unacceptable, so why is it increasing instead of decreasing?

One likely reason is that drivers are being distracted by mobile phones, satnav and music. How are drivers being trained to interact with these distracting systems when operating their vehicles? The obvious answer is that no training is required or legislated. I always joke that pilots are the only people trained to talk while operating their machines.

Car manufacturers are striving to offer automated information systems to be used without any specialised training. The vehicles' handbooks advise drivers against interacting with these systems while on the road, but who reads those manuals anyway? Perhaps their warnings aren't enough to deter drivers from playing around with their infotainment when they should be concentrating on driving their cars.

The suite of vehicle automation has been steadily increasing; adaptive cruise control, lane assist, collision avoidance and stability augmentation to name but a few. They enhance vehicle safety. The addition of semi-autonomous systems like 'autopilot', however, maybe generating a false sense of invincibility in drivers' minds. In fact, they may just be providing the tools for enhanced bad driving.

The overconfident mentality of many young pilots appears to be shared by many car owners who have these systems installed. But, unlike pilots, drivers have little or no practical instructions on their operation.

In the past few years there have been four high-profile fatal accidents involving self-driving cars in the US. There's a recurring theme: the sensor technology doesn't guarantee an appropriate response from the installed algorithms to avert a collision. The human operator has neither the training nor the awareness to take over manually when a situation requires rapid action.

The sensor–algorithm conundrum present in aviation is becoming a prominent black hole in the design of autonomous vehicles. This technology places drivers in a No Man's Land when automation fails and it's too late for them to save the situation. Human operators are the last line of defence to address complex automation failures – and the first in line to be blamed if they can't manage them.

In March 2018 a pedestrian, 49-year-old Elaine Herzberg, was struck and killed by an autonomous Uber taxi in Arizona. In the wake of this tragic accident, the CEO of Toyota, North America, told a journalist, 'The reality is there will be mistakes along the way. A hundred or five hundred or a thousand people could lose their lives in accidents like we've seen in Arizona.' He estimated that autonomous vehicles could eventually save 35,000 lives in the US annually: 98 per cent of deadly crashes are blamed on driver error. If hundreds may die so that thousands may live, perhaps a 'martyr's clause' should be included with each autonomous vehicle; this would guarantee the payment of compensation when a death or injury exposes a flaw that the manufacturer can now rectify. It seems we're signing up as part of manufacturers' research-and-development teams to fine-tune the design and safe operation of their products.

But where does the liability for automation-related accidents lie: with the manufacturer, the software engineers outsourced to provide the algorithms, or the owners or operators? In the heavily regulated world of aviation, laws in every country identify the pilot-in-command as ultimately responsible for any accident. There's a history of legal cases where pilots were found culpable because they didn't use their automation, in the form of the autopilot, and an accident occurred.

Human operators of automated systems are positioned to absorb the blame when an accident occurs, in what's known as a 'moral crumple zone'. Just as the crumple zone of a car is designed to absorb the force of impact in a crash, a human operating a highly complex automated system may become simply a component that bears the brunt of moral and legal responsibilities when the system malfunctions. A widely accepted notion is that the autopilot and associated automation are smart enough to save the human every time. It isn't thought possible that the software and automation could both fail. It seems the operator's central role becomes that of soaking up fault, even if they only had partial control of the automated system.

Western legal systems are structured to apportion blame, and laws must evolve along with the technological landscape. Locating liability and accountability must be expanded beyond the immediate physical system of human and machine to include the actions of designers and engineers who create the software of automated and autonomous systems.

How can this technology be allowed out into the community when its performance is questionable? What if its manipulated and an adverse, unintended consequence is the result? And is the push for highly automated systems coming from governments or from the competing tech companies?

As Earl Wiener stated, 'Exotic devices create exotic problems,' and he questioned the assumption that automation can eliminate human error. So do I. Fallible humans are writing the software and building the automated machines, so they're predestined to fail in some way. The automation doesn't eliminate human error, but it does create new and unexpected opportunities for exotic failures.

In April 2018, a Tesla driver in the UK was filmed by other motorists cruising down the highway while lounging in the passenger seat. This driver had put himself in the hands of his car's autopilot function. Fortunately, local police reacted swiftly and the driver was apprehended, fined and disqualified for eighteen months. I guess he did it because the technology provided him with the tool to do it.

Can automation ever be infallibly safe?

The recent autonomous vehicle and aircraft accidents are prime examples of the new age of exotic failures that we're racing to embrace. Societies, companies and governments must decide how much automation is worth the risk.

# 29.

As 2012 dawns, I'm still stuck flying the A330. The ATSB accident report into QF72 doesn't allay my concerns — I distrust the Airbus and its fly-by-wire operation. I'm still committed to flying as a commercial pilot but I'm not enjoying myself.

I'm dreading each flight; I'm having trouble separating myself from the security of my home to go out and make a living. Even though I pack the same clothes for practically every trip, I keep leaving that task to the last minute. Hearing the rasp of the zipper as I close my luggage is like the crack of the starter's pistol.

I'm not the Captain Kev of old. The memories of the mangled mess inside the aircraft haunt me as I do everything in my power to prevent that from happening again.

A ten-day trip from Sydney to New York in April 2012 shakes my resolve to persevere. It's pockmarked with a variety of challenges mixed with the debilitating exposure to turbulence and a potential loss-of-control through false activation of the aircraft's protection modes. My ability to stay in the cockpit is under threat again.

I'm approaching the four-year mark since the accident, and my situation is worsening. I decide to seek professional assistance again. I've avoided medication so far, but maybe that's the next step. I contact the welfare officer in my pilots' association to ask for his recommendations, and he gives me a name of a psychologist. I make an appointment. I'm about to enter new territory.

At the same time, I make the professional decision to stop flying again. I start to attend appointments with a psychologist recommended by our pilots' association. He confirms that I'm psychologically unfit to fly. But there's a catch. 'As a psychologist I can't prescribe medication,' he explains. 'Only a psychiatrist is qualified for that.' I stop seeing this psychologist and seek another specialist.

My treating doctor refers me to a psychiatric specialist who's already providing assessments to CASA for the annual renewal of my flying licence. This specialist recommends the introduction of medication, in the form of a selective serotonin reuptake inhibitor (SSRI), to address my mood and body response issues. SSRIs are used to assist sufferers of depression and anxiety-based disorders. They chemically alter the brain's neurotransmitters through increasing levels of serotonin, but they're simply described as 'blockers'. These are powerful drugs that can have debilitating side effects, and it takes some time to find the correct dosage.

I wonder if I'll ever return to the life I built before that flight. My Maslow's hierarchy of needs was hit by a MOAB – Mother of All Bombs – and I'm operating in the fractured remains of the lowest two levels.

People close to me know something isn't right, but I look whole from the outside. I turn to analogies to describe my feelings, my realties and my frustrations. But the technical failures of QF72 and the complexities of PTSD are beyond the comprehension of the majority of my intelligent and experienced friends, through no fault of their own. It's frustrating and exhausting to try and describe what I'm feeling to someone who's never experienced the aftermath of a traumatic event. I didn't grasp the fallout until it happened to me. I feel there is no one I know who can help with my unique difficulties. I am compelled to inform my company that I am not okay.

*

In the navy we would always joke, 'There's no such thing as a free lunch.' This cynical reality takes on new meaning for me during this period of uncertainty.

Eventually, after extensive consultations with my specialist, I confirm my intentions to return to work in September 2012.

I visit the Qantas medical department and give them more background on the past four years. They'll provide assessments that will contribute to the reissue of my flying licence from CASA and my requalification programme. I ask their opinion, based on my body response, nightmares and disrupted sleep patterns: do I have PTSD?

They agree that I present with criteria for PTSD, and quote DSM-5.

DSM-5 … What's that?

I continue with my own research, educating myself about the injury that could potentially ruin my life.

*The Diagnostic and Statistical Manual of Mental Disorders, Fifth Edition* (DSM-5) is a diagnostic tool published by the American Psychiatric Association. In 2013, Version 5 was published with specific changes to the testing and diagnosis of PTSD. There's no better organisation to quote regarding PTSD than the US Department of Veteran's Affairs and their National Center for PTSD. The DSM-5, used by Veteran's Affairs, provides a checklist of specific symptoms; all must be present to confirm the presence of the disorder.

I tick practically every box. I was exposed to death or serious injury. I'm continuing to relive the traumatic experience through memories, nightmares and the stimuli of flying. My sleep is adversely affected. I'm isolating myself and not enjoying activities that I did previously. I'm easily irritated, and my boundaries of patience and frustration are easily breeched. I'm hyper-vigilant and on edge when I go to work. My DSM-5 scoreboard is pretty full.

I didn't get a manual on PTSD after the fact, so all of this is a revelation. My GP provided a list of potential outcomes but it is up to me to write a game plan specifically for my situation. The stark reality that my career longevity might be adversely affected is a bitter pill to swallow. Each year I assess my position; am I pushing it too much to stay in the air?

But I know how to 'push it'. After all, pushing it is part of a fighter pilot's job description. I'd had to do it countless times. I recall, on my second deployment aboard USS *Enterprise*, getting the opportunity to test my limits. This flight was probably my most enjoyable and also the shortest of my carrier career.

I was a certified post-maintenance test pilot, which means I was authorised to take up repaired F14s and ensure they were completely functional before being cleared for operational use. On this day, I was asked to take up a jet to perform a simple angle-of-attack calibration check.

This check simply verifies that the indicated angle of attack displayed to the pilot is correct for the aircraft's weight and configuration. It doesn't take long to complete and our squadron needed the aircraft to meet the day's flying schedule requirements.

I was required to perform this routine check during the timeframe of a launch and recovery cycle. The cycle involves launching a group of aircraft first, then recovering the aircraft that launched two hours prior. At the beginning of a deployment, that launch and recover evolution would take about fifteen minutes; towards the end of the cruise, about eight minutes.

I wouldn't have much time to complete my task, and it was further complicated by the amount of fuel on board. The tanks were full – 20,000 pounds full. Even with full afterburner, I would have trouble burning and dumping enough fuel to land at the maximum permissible weight in this short time.

Fighter pilots normally sweat their dwindling fuel load, but I had way too much for this flight. So I would have to improvise and make it happen.

Our squadron's flight deck chief had pulled every string to ensure I launched off the catapult first. In no time, I was connected to the catapult and strapped in for my wild ride.

The catapult did its job and punched me off the deck. I kept the afterburners lit and accelerated to 540 knots, which was the maximum speed with external fuel tanks fitted; the ones loaded to the brim with fuel.

Safely outside of the carrier's landing zone, I selected full afterburner and rapidly climbed to 10,000 feet to do my check. I was there in about a minute.

Meanwhile, the launch cycle was in full swing and the recovery aircraft were circling the ship like a swarm of bees returning to the hive.

I had to work fast. With speed brakes out, I quickly slowed down to landing configuration with gear and flaps out. My RIO recorded the angle of attack for the various indications and our check was successfully completed in a few minutes. Now I had to get rid of enough fuel to land.

The fuel dump system works best if the engines are at a higher power setting. It is prohibited to dump fuel with the afterburners lit, unlike the dump-and-burn an F111 can do. Someone found out the hard way that doing this in an F14 is a no-no.

I was at 10,000 feet with 8000 pounds of fuel to burn. And the landing recovery below had commenced.

I selected full power, without afterburner, and initiated fuel dumping. I corkscrewed slowly down towards the ship using maximum g turns to help the fuel dump do its job. It was agonising to watch my fuel gauges slowly decreasing through the haze of a constant +6.5 g turn.

It was taking too long and I was running late. The recovery was complete and all eyes on the flight deck were directed skyward now at my lone F14, which, with a white cloud of fuel vapour pouring out of the tail, no doubt looked like a novice skywriter.

The *Enterprise* was churning through the ocean, its long wake signalling to me that I better pull my finger out and get aboard. But I was still too heavy. And now I was about to join the landing pattern and had no altitude to play with.

I secured the fuel dump as I descended to 600 feet behind the ship. My only method available now to decrease my weight was the afterburner. I lit the fires and advanced the burners to full Zone 5. Manually I swept the wings back and slammed my tailhook down as my F14 accelerated like a bat out of hell towards the wake of the Big E.

For a normal recovery, an F14 would join the landing pattern at around 350 knots, extending upwind of the ship for 45 seconds before breaking into the landing pattern. This gave the pilot ample time to prepare the aircraft for landing. You don't want to rush something as potentially dangerous as a carrier landing.

I didn't have time for this type of normal approach; I had made the ship wait long enough. My speed increased to 540 knots quickly as I pointed the nose of my jet at the back of the ship. Through my mirrors, I could see intermittent clouds of water vapour flash over my wings. I was going fast and the air pressure was compressing the moisture in the air.

It must have looked ominous from the deck of the *Enterprise* as this F14 thundered towards the fantail of the ship in full afterburner, the wings enveloped in cloud. I had my aircraft on its side in 80 degrees of left bank as I crossed the wake.

I increased the bank to 90 degrees and pulled to 6.5 g as I passed the stern. I kept turning hard, my nose and canopy the only parts visible as the F14 was enveloped in vapour.

Could I pull this off? I hadn't explored this extreme edge of my capability before. Well, if you're gonna screw up, at least look good.

I kept the wings in full aft sweep as I used the aerodynamic drag of the wing and maximum g to wash off speed. I continued to turn hard and line up for landing as I sweated my speed versus my ever-decreasing distance to the ship.

At 250 knots, I threw the landing gear down. Passing 225 knots, I selected the wing sweep to auto, allowing them to sweep forward, and selected full flap. I was still in 45 degrees of bank and rapidly approaching the wake with the engines at idle power.

Rolling out on final approach, I was still fast but also displaced further back from the ship than normal. My RIO called the ball and our approach was completed without incident.

I was buzzing with adrenalin as I made my way to our ready room. What a ride!

As it turned out, my superiors weren't so excited. It was the Cold War, and we were operating in close proximity to the eastern islands of the Soviet Union. The day before, a Soviet attack submarine had been photographed at periscope depth in the wake behind *Enterprise*.

Sometimes you have to take a calculated risk and throw all your chips on the table. You've got to push it. Will pushing it now help me to defeat PTSD?

My research helps me to understand my frustrations, accept the emergence of my symptoms and become aware of the physiological outcomes associated with PTSD: the 'psycho-biological' part.

The body responses I'm experiencing outside of flying, such as embarrassment and frustration through social interaction, are unmuffled. It's as if someone has taken an iron bar to my exhaust system, knocking out all the baffles and removing all the inhibitors to my fight-or-flight chemical production. My system is being flooded with cortisol, the main stress hormone. I'm getting free-flow for the smallest stimuli.

These realities have taken time to expose themselves, and it's apparent how easily they can remain hidden, as long as the mind is busy and distracted. My reasons for stopping to analyse my situation again are now justified and critically important to my own self-awareness. My research also confirms that the medication may help improve my quality of life and possibly preserve my career. It is a painfully slow journey of discovery.

Am I just supposed to suck it up, be a 'tough guy' and suffer in silence? No, I won't remain quiet. But ultimately, it's up to me to sink or swim.

PTSD drops a sufferer into an emotional minefield without a map; you either remain stationary and hope for the best, or try to navigate your way out. Every day, I have to figure out a new way to navigate out of danger while maintaining my shredded self-esteem and security.

*

In March 2014, Malaysian Airlines flight MH370, a Boeing 777, disappears without a trace en route to Beijing. Was it pilot suicide or an automation failure?

Soon after, in April, I'm operating QF24 from Bangkok to Sydney. The flight encounters an extensive area of thunderstorm as we cross the Australian coast near Darwin. We spend the next ninety minutes dodging weather and enduring turbulence. We've been cruising at a reduced speed with the seatbelt sign on for most of that time. Our passengers sleep in the darkened cabin.

As we pass abeam of Alice Springs on our southeasterly course to Sydney, the skies clear and the turbulence subsides. We reset our cruise speed, and I notice the tailwind at our altitude is almost directly behind us at 180 knots. We're scooting along nicely now. A Thai Airways 747 is behind us, 2000 feet below and on the same track to Sydney.

I hand control over to the first officer and leave my seat to retrieve the aircraft's technical log from its stowage, planning to make a routine entry. I place it on the glareshield and take a short break to stretch in the area behind our seats.

Abruptly, the turbulence starts again.

As I scramble back to my seat, my first officer turns on the seatbelt sign while manually reducing the speed for turbulence penetration. Not even five seconds has elapsed.

My instrument scan reveals the makings of a jet upset: my aircraft has been upset by the changing environmental conditions and is deviating from its stabilised flight path. My instruments reveal indicated speed at maximum, altitude 39,300 feet and increasing, vertical speed climbing rapidly, and

Autopilot 1 engaged. Autopilot *still* engaged? It should either hold altitude or disconnect itself. As I digest this snapshot of the jet's extreme performance profile, a further scan of the navigation display indicates a headwind has developed very rapidly.

I have to make a decision within the next second. Do I leave the autopilot engaged and let it manage the jet upset – probably propelling everyone into the ceiling as it pushes the aircraft back down to 39,000 feet – or do I disengage the autopilot and manually arrest the trajectory to prevent the aircraft from pitching down to its commanded altitude?

I disengage the autopilot to stop the climbing trajectory. My trusty first officer extends the speed brakes on his own initiative. The aircraft porpoises aggressively around 39,300 feet despite my manual intervention, abruptly climbing and descending in vertical speed and I notice on my navigation display that the Thai Airways aircraft has departed its altitude by 400 feet and is still climbing behind us.

My Airbus is in a washing machine of turbulent air as I fight with small sidestick inputs to maintain a constant nose attitude. The up-and-down shifts in vertical speed make it feel like we're on a roller-coaster, reminding me of the first QF72 pitch-down.

It's about 4 a.m. in Sydney, not the best body-clock time to be dealing with a jet upset on a pitch-black night. Maintaining a constant nose attitude on my display is almost impossible.

We abruptly leave the turbulent air. It seems like minutes have elapsed since I strapped back in, but it might have only be a minute. I stabilise the aircraft's nose attitude, then re-engage the automation to regain our assigned altitude of 39,000 feet. I report severe turbulence and windshear to air traffic control, then I take a deep breath. My heart is beating hard, but not near the level it was on QF72.

I give it a few seconds, then call the rear galley for a status report.

'A bit bumpy,' is the nonchalant response from my flight attendant.

I breathe a big sigh of relief.

This was a 'classic' jet upset caused by the environment, not the computer systems. We've dodged a bullet, but will I be nimble enough to avoid other bullets in the future?

# 30.

Over the next few months, there are three more significant Airbus accidents. The most significant involves pilot mental health.

The title of the report is chilling: 'Deliberate Flight Into Terrain'.

Germanwings Flight 9525, an Airbus A320, was a scheduled service from Barcelona to Dusseldorf with 144 passengers and six crew. The first officer waited until the captain left the flight deck to use the toilet, then locked the cockpit door and deliberately flew the aircraft into the French Alps.

This accident is incredibly disturbing to me. There is a history in commercial aviation of pilot suicide events while on duty and here is another to add to the tragic list.

Unfortunately, this young pilot was battling depression and found a way to consult with different specialists and receive different medications from each one. He was declared unfit to fly by each specialist but failed to share those restrictions with his airline management. Then he abused the medications, mixing them without supervision. The result was his drastic conclusion to end his life with the use of an aircraft.

Automation can't protect against a suicidal pilot.

The BEA report on this accident recommends further scrutiny of pilots with mental-health issues and promotes pilot support programmes.

I read an article that intimates there's widespread denial in the airline industry about pilot mental health being an issue. Pilots tend to under-report their health, in general, and the only mechanism to identify issues is the personal responsibility of the individual pilot to self-report. A former Lufthansa official offers that it's difficult to screen pilots who are determined to keep flying and are willing to lie in order to do so. If a pilot chooses to hide problems and symptoms, there are no existing tests to determine whether that pilot is healthy and fit to fly. Two doctors who were formerly employed at Lufthansa express their astonishment that pilots with mild or fading depressive disorders are taking antidepressants without the knowledge of flight doctors – and continuing to fly before and after the Germanwings tragedy.

This is a serious problem. The Germanwings accident should never become a reality for any airline if they truly value their employees.

I decided long ago to be responsible and mature in my management of PTSD, regardless of the negative impact it may have on my career longevity. But not every pilot is willing or able to do the same; this young Germanwings pilot decided to hide his situation and continue flying.

*

In April 2015, I'm challenged again, this time while operating a domestic flight from Perth to Sydney.

We're flying a daylight sector and the weather is clear. The forecast shows a disturbed jet stream in the vicinity of our descent point, but our flight plan and weather package aren't predicting turbulence. I have an experienced first officer and cabin crew onboard, and our passenger load is full.

Cruising at 41,000 feet and in smooth conditions, I notice solid cloud cover below, obscuring the ground. There's no thunderstorm activity visible or displayed on our weather radar. Just in case, I let the CSM know of the potential for turbulent conditions, and plan to prepare the cabin early for our arrival in case we encounter turbulence on descent. We haven't received any pilot reports of turbulence from air traffic control, but I want to minimise any risk of injuries.

On descent, the air is still smooth. When the plane encounters the cloud tops around 30,000 feet, we announce, 'Passengers and crew, return to seats and fasten seatbelts', and we illuminate the seatbelt sign. Crew have one minute to secure themselves.

Two minutes later, passing 26,000 feet, we encounter severe turbulence. 'Severe' means I have difficulty reading my instruments and the aircraft is moving violently. I use one hand on the glareshield to secure my position.

'Kev, this is severe turbulence!' the first officer remarks.

That's his prompt for me to reduce the speed for turbulence penetration; my hand is already winding the speed back.

This is the worst turbulence of my career, and I can't stop worrying about injuries. I hope the passengers have listened to our warnings and instructions.

Passing 20,000 feet, we're in the clear. I tell air traffic control of the severe turbulence and the affected altitude band. A following Qantas aircraft hears our report and adjusts their descent to stay safe.

We all keep our seatbelts on for the rest of the flight, and I check with the cabin crew for any problems. They advise of no injuries but that many passengers were vomiting. They also advise that they weren't able to stand or walk during the turbulence as they rushed to their seats. This last comment concerns me; I hope they were secure in their seats as warned.

After a routine landing, I wait by the front door to watch my passengers leave. Many are shaken but thank me for delivering them safely. I can smell the vomit tainting the air.

I approach the assembled crew for a company-mandated debrief of our turbulence encounter. As I join them, the CSM has already commenced her debrief. She reminds them of the requirement to be seated within one minute after our seatbelt announcement.

So it seems they were not seated. They had delayed securing themselves, and my worst fear of passenger and crew injuries was just narrowly averted. My stomach drops.

I step forward to offer some stern advice. 'Are you nuts? If you're injured at work then you become a liability. You cannot take these risks and expect to come out the other side in one piece. Take it from me, you place your health and career at risk by making your own turbulence assessment. Don't ever do this again.'

My blood pressure is in the red zone as I walk off the aircraft. My worse fear almost became a reality.

As always, it's up to me to self-assess and self-report if required. I'm deflating but still persevering; my performance is still at a high level. More damage is done in the aftermath of these types of extreme events, though. My increased vigilance and ultra-conservative approach have prevented passenger injuries again, but I sense that the precarious ledge I am navigating is narrowing.

*

On 31 December 2015, I'm scheduled to fly a six-day trip to Hong Kong, and the first stop is a night landing into Perth.

Severe turbulence is forecast below 5000 feet, along with high temperatures and gusty winds. These winds roll over the foothills to the east of the airport, becoming disturbed and unpredictable by the time they blow over the runway. The adverse conditions are confirmed as we check the weather before our descent.

I've uploaded additional fuel in Sydney to cater for several approaches and a diversion to Adelaide. Perth is an isolated airport and Adelaide is the nearest divert field, three hours' flight time away, and unaffected by the gusty wind conditions.

Our landing is the first officer's sector, but I decide to do this one myself. He doesn't seem too upset about this. The A330 can be a handful landing in these types of conditions, and the nuances of the fly-by-wire system can be counterintuitive with the control management required to maintain an appropriate flight path to landing.

Also in preparation for rough conditions, we brief and prepare our cabin crew and passengers. As we approach the airport from the east, we hit the turbulent air. I allow the autopilot to intercept the runway centreline but am forced to disconnect the autothrust soon after. The wind makes the power surge against the wind gusts. I then disconnect the autopilot to stop the plane's wild porpoising as it compensates for the turbulent air.

I manipulate the power manually to maintain my speed, energy and our vertical path towards the runway – but the gusts are severe. A few times I'm near minimum and maximum power to control the approach speed.

I'm on guard. As we cross the runway boundary, the nose is weather-cocked, pointing into the gusty wind like a sailboat on tack, but the aircraft is aligned with the centreline stripe and the engine power is maintaining my energy. I wait until we're 20 feet above the runaway to kick the aircraft straight with the rudder, leaving the power on as I raise the nose to flare and wait for the wheels to touch before I reduce the engine thrust to idle.

We rock and roll down the runway, the aircraft decelerating while its wings are buffeted by the wind.

My first officer tells me he was grateful to be flying with me in those gusty conditions and in manual mode. I just shake my head and shrug. I was confident in my manual flying skills, as always. But it was the most turbulent, difficult approach I'd flown in my Airbus career.

Through the cockpit window we watch as an aircraft executes a go-around from an aborted landing.

My passengers and crew are all very complimentary and happy to be in one piece as I stand by the door and wish them a happy 2016. I think about holding my hat out to receive tips, but I just wave and say, 'Happy New Year.'

Happy New Year, Captain Kev. I take several hours to power down for a fitful sleep at the hotel.

The next night we're flying the red-eye service to Brisbane. The gusty winds are still blowing at Perth Airport, and windshear conditions are reported by an arriving aircraft prior to our take-off. 'Windshear' means abrupt changes in the prevailing wind's velocity and direction. I inform my less-experienced first officer that I'll be the pilot-in-command for the take-off. Windshear is an abnormal flight condition and potentially a serious risk.

As we taxi out to the runway, I brief my crew and passengers about the rough conditions. Our take-off procedure requires maximum thrust and, coupled with the strong winds, make for a space shuttle-like lift-off. I must lift the nose higher than normal to control our rapidly increasing speed and this allows for a quick exit from the turbulence. I hand over to the first officer and we set course towards Brisbane and the waiting sunrise. We arrive at dawn without further incident.

I arrive at our hotel, exhausted. Can't I just have some uneventful flights?

The next morning, our crew meet at Brisbane airport for our departure to Hong Kong. There are three pilots for this flight, and together we look through all the usual paperwork. The weather forecast is fine but I stipulate an additional hour of fuel to cover any operational surprises. My pilots are happy to approve this, not wanting to manage the risk of a potential diversion.

Several hours later, we depart Australian airspace. Our flight progress has been smooth and routine.

*Ding!*

No, it isn't the master caution chime this time – it's a message from Qantas Operations Control: 'FLIGHT WATCH ALERT'. There's un-forecast low cloud and reduced visibility at Hong Kong Airport for our arrival. We'll need an additional sixty minutes of contingency fuel to legally continue. Do we have enough?

I nonchalantly tear the message out of the printer and pass it to my wide-eyed first officer. His fingers move quickly to the keys on his navigation computer. I already know the answer but it's good experience for him to check it himself. We have the required fuel, so I send back our response: 'Affirmative.'

'Nice fuel order, Captain Kev,' I remark as I take a long drag from my imaginary cigarette. Without my extra order, we would have made an unscheduled stop in Manila.

The low visibility means a potential automatic landing into Hong Kong. As we approach our descent point, I take over as the flying pilot to complete an automatic landing. The automation does a good job helping us land in these greasy conditions. Aircraft are parked on taxiways everywhere around the airport; the adverse weather is generating huge delays.

I walk slowly to our waiting crew transport, feeling again like Iron Mike Tyson has me on the ropes after round three.

In isolation these challenges would be relatively routine, but my PTSD complications can make them feel debilitating.

The following night we head back to Brisbane. Again, I order extra fuel because nothing goes according to plan with me these days.

It's a remarkably clear night as we overfly a brightly lit Hong Kong and set course for Brisbane. Entering airspace controlled by Manila, we're interrupted by a loud *ding*: another 'FLIGHT WATCH ALERT'. This one has a more sinister message. A volcano has erupted, affecting our route with an active ash cloud above 40,000 feet drifting towards our track. To avoid the eruption, we'll need to follow an amended flight plan. Do we have enough fuel for the new route?

The first officer and I load the revised route and check the fuel. The new routing clears the volcano by about 40 miles, but it's too close for my comfort. I look at an alternate route providing 100 miles of clearance; we have enough fuel to cover that modification, too.

Once we have approval from Australia and Manila, we program our diversion and alter course. From my flight library I pull out the *Quick Reference Handbook* and open it to 'Volcanic Ash Encounter', then to 'All Engine Flameout'. (Volcanic ash is a serious threat; Google 'volcanic ash events for British Airways and KLM'.) I discuss contingencies with the first officer as we approach the volcano, isolated safely off our right wingtip.

While we transit the volcanic ash area, I'm in a high state of readiness, even though we're comfortably upwind. Fate hasn't been that kind to me of late, and I'm relieved when we pass through without incident.

Our flight is completed without any more distractions or surprises, but once again, I felt tired from absorbing the stresses presented during these six days. My reservoir of resolve had sprung some leaks. Outwardly, I was calm and cool, but I was bleeding energy.

I have survived the most difficult trip of my career; severe turbulence, low visibility approach, volcanic eruption and potential flight diversions. The aftermath is exaggerated by my increasing difficulties coping with the daily challenges of PTSD. I'm really starting to feel the strain by now and I know something has to change.

I wonder if my next trip will be the one where I become the cornered boxer and get pummelled to the canvas by something outside of my experience to handle. The body blows are starting to land and I'm feeling like I'm on the ropes.

Maybe I shouldn't step into the ring again.

# 31.

Throughout most of my life, I never really thought about the end of anything I started. As a child building plastic aeroplane models, I couldn't wait to finish one so I could start another. Every flight in my career had a start and a finish, but the end of one was only the beginning of another. The thought of my flying career ending was something I never contemplated, even though I knew it would happen one day. I would address that day when it came – and I'd be the one who ended it, on my terms if possible.

Flying fast jets is a young pilot's game, as much as the egos of fighter pilots profess otherwise. Circumstance, lifestyle change and health are some of the variables that affect that conclusion. For me, it was a lifestyle change, and I accepted that conclusion gracefully and realistically.

In 1986, I was flying the Mirage 3 with the Royal Australian Air Force in my role as a US Navy exchange officer. I was married and had a daughter, and I was due to rotate back to sea duty in the navy at the end of this assignment. I chose to finish up my military career and not accept orders to a new squadron, which was destined to return to sea. I was a family man now – returning to sea duty and the stresses it can place on relationships wasn't supportive of a stable family life.

I was a highly qualified and experienced fighter pilot, on track for an illustrious military career, and my country was still involved in the Cold War. Would the fight be weakened if I left the service? I agonised over this. But I realised that no one is indispensable, as much as we like to think we are. In the end, though it wasn't an easy decision to resign, I felt it was the right one.

Flying a high-performance jet is a melding of human and machine. The aircraft and its performance become an extension of your being, like riding a Ducati motorcycle on steroids but in three dimensions. Separating yourself from this dependence is never an easy decision or process, but it must happen for each jet pilot at some point in their careers. When that day arrives, transitioning back into an earthbound mortal is nothing to look forward to.

My last flight would be a one-versus-one training mission for a young Aussie pilot doing the Mirage conversion course.

The flight was uneventful. The student pilot was aggressive and displayed excellent handling of his jet and in the strategies he used to defeat his 'bandit' adversary. I made him work for it, while he effortlessly fought off my attacks according to the mission profile.

I had briefed my wingman that we would return to base separately; I wanted to say goodbye in the right way to my squadron mates and to the fighter base where I'd flown for the previous three years. I was the first US Navy Exchange Pilot to the RAAF and my departure needed to be commensurate with that honour.

The Mirage is a magnificent machine at low level. The delta wing slices through the air effortlessly while the afterburner provides the potential for incredible speed. Buildings, trees, hills streak past the canopy in a blur while those unsuspecting on the ground don't hear you're coming until you blast past, adding to the pure exhilaration of flying a fast jet.

I would miss this part of military flying enormously. The RAAF pilots were excellent low-level hoons; I would show them what I learned from flying with them.

After our last sequence, my student rejoined in close formation on my right wing. I gave him a thumbs-up and a smart salute, then pointed to him with two fingers to indicate he was now the lead pilot. He nodded and I broke away hard to the left, generating some distance between us for the return to base.

As I descended through the initial point for the runway, I eased up the power and started to accelerate. I continued my descent to 100 feet above the beautiful shoreline of 8 Mile Beach near RAAF Base Williamtown, as I pointed directly at my squadron's admin building.

My indicated speed was approaching 620 knots. I was still subsonic but pushing a lot of air with my Mirage, and the blue glow of the afterburner kept my speed high. Passing the perimeter fence, I dropped a few more feet to fly directly over the squadron building and the flight line of parked jets. I was moving fast and caught a glimpse of my squadron mates assembled near the flight line as I zorched past.

The many windows of the base buildings flexed as my pressure wave passed over them – thankfully, they didn't break. The duty officer perched in the small control room announced, 'That was Sully', over the PA system in a laconic Aussie twang when I thumped the building.

It had been an eventful nine years since I performed my mini airshow over my frat house in a Cessna 150. This was a different display.

I slammed the throttle to idle, pulled up hard and snapped the jet into a hard left bank while applying maximum g to enter the landing pattern. My aircraft was enveloped in a cloud of 'ectoplasm': pilot slang for describing moisture in the air being compressed by pressure change and g-force, forming a vapour cloud around the jet.

As I rolled out on final approach, I eased up the power to maintain my speed until the runway threshold, then slid the power back and touched down. Emulating a top-fuel dragster, I deployed the drag chute and slowly decelerated to taxi speed with my chute billowing behind.

Exiting the runway, I cracked the canopy open and welcomed the fresh air into my tight cockpit. Sitting in the Mirage, I always felt like I was in a Formula 1 car, but this car had an afterburner. I would miss this extension of my being, this instrument of my passion. My return to the parking tarmac was deliberately slow; I knew this moment was not to be repeated, and I wanted to savour it.

My Aussie squadron mates were standing by with a firehose, primed and ready. They were all busy preparing for their flights, but the squadron had stopped for these few precious moments to commemorate the end of my military career.

Safely parked, I shut down the engine and hesitantly eased up the canopy. A base photographer followed the ground crew to my jet as I stared thoughtfully forward. I had survived nine years of dangerous high-performance flying, two carrier deployments and all the pressures connected to this extreme life. I was thankful and paid silent respect to my friends who had not been so fortunate. The photographer's camera flashed at this moment, capturing my slight smile and pensive stare.

I climbed out of my Mirage and down the ladder to the tarmac, my old VF-114 helmet propped on the canopy railing. The firehose was charged and the water pressure almost knocked me over as I reached down and gripped my life vest inflation toggles, feigning that I would activate them to save myself from this dousing. My squadron mates were enjoying the spectacle of the Yank getting blasted, but there was respect and comradery flashing in their eyes.

My tour proved influential and a new RAAF was emerging as the F-18 Hornets began to arrive at the base. I felt proud to be part of that transformation.

My mates presented me with a bottle of champagne, and a few chugs of bubbly was my final act as a fighter pilot. As I walked off the flight line for the last time, I reviewed my military career. I'd flown in the forces of two countries and in 6 different aircraft types. I'd flown as low as 10 feet and as high as 55,000 feet. I've seen the curvature of the earth. I'd flown at zero airspeed and faster than twice the speed of sound. I survived 350 carrier landings without mishap. I was a Top Gun graduate. And I'd lost a lot of friends but I was able to walk away.

*Yeah*, I was thinking, *that's good enough.*

Dripping wet, I walked away from this exciting life and into the world of commercial aviation.

*

In January 2016, I'm burdened after my last trip and mired in a month-long cycle of soul-searching. My aviation medical is due for renewal in February.

I have a vivid dream.

I'm flying at night; a supervised flight with a senior pilot observing. He puts me to the test, and I pass all the requirements. For some reason, I look down through the cockpit window and see a long road in an isolated area. At the end of the road is a car with its headlights on. I put down the wheels and flaps and land on that road, taxiing towards the waiting headlights. I shut down the engines and calmly climb from the aircraft on a ladder. The other pilots only stare in silence and don't try to stop me.

My daughter is beside the car. She waves and smiles as I approach. I give her a big hug and together we drive away. I don't look back.

I wake up with the realisation that I have to make a choice. I've been labelled a PTSD sufferer, but what does that really mean to me? I have to strip away the emotion, exposing the realities of persevering or ending my flying career.

Flying is my passion. I've dedicated my life to being a pilot, and not an ordinary one. Following the course to fly fighters off aircraft carriers required total commitment, superior performance and courage. I cherished every second I was strapped into those fire-breathing machines. The good dreams I have are centred on that exciting and dangerous life.

Now, I've survived something that science-fiction stories are made of. The Kevin Sullivan I invested all my time and resources to develop has been compromised, battered and thrown into a minefield without a map.

I've become very isolated – when you've been to the Moon, you can only talk to astronauts. I'm a self-exiled cave man, which I've adopted as a coping mechanism. If I stop flying, I'll lose the security and structure of employment, but should I stay tethered to a workplace that will only complicate my life further?

The boundaries between PTSD and depression are blurred. I've resorted to medication to prolong my career, but the side effects have left me numb and robotic. For me, the medication does not prevent feeling worried or threatened, they only stop the body's untethered release of chemicals in response to these feelings. Is this how I want to live my remaining years?

I have to face the truth: my passion has become a cross to bear.

This life-changing event has also damaged my daughter. She's had to watch helplessly as my persona changed. She knows she could have lost her dad on this day and that is very confronting. It's okay honey, I'm still here.

In my dream she was there to take me away, waiting for me with a big smile and relief in her sparkling eyes. She knew I had decided to stop, and she was happy.

That dream answers all my concerns, and I listen to it.

I meet with my specialist to discuss my position. He's been monitoring my struggle, so my news is no surprise to him. He agrees it's time for me to stop.

However, when I meet with Qantas medical, they seem a bit surprised.

'I'm not okay,' I repeat. 'And if I truthfully answer the mental-health questions in my upcoming aviation medical, I won't pass. What do you want me to do? Take the medical and fail, or just stop now?'

'But you're back flying.'

'Yes, but that doesn't mean I'm okay. I don't think I'll ever be okay. PTSD doesn't just go away. The company policies encourage us to self-report and … well, I just have.'

A month later, I'm back. I inform Qantas medical that I'm finished taking medication for my condition and that redeployment within the company isn't an option.

My decision is irreversible. I've honoured my commitment to stop flying if I ever doubted my ability to continue. I made the courageous decision before it was made for me. I've been trying to navigate my way out of No Man's Land, back to safety and salvation. If I need to leave something behind so I can move forward, then so be it.

*

In early July 2016, I am medically terminated. Medical termination acknowledges the serious nature of my injury and the catastrophic effects that PTSD can inflict on an employee.

On my last day at headquarters, I pose for a couple of photos with my close colleague, Captain Matt Hicks. He was the hero first officer on QF32, which suffered a severe engine failure in 2010. We exchange a quick handshake, and that's my departure, done and dusted. I keep searching the ceiling for the balloons to fall and the confetti to be released, but it's not to be.

I exit through the sliding doors and don't look back. As I walk to my car, I summarise my commercial career.

I'd flown three different aircraft types for a total of 16,000 flight hours. I'd flown millions of miles and carried hundreds of thousands of passengers. I always delivered them safely to earth. I had been tested many times with equipment failure and environmental extremes and persevered. I had managed and survived what could have been one of the worst aviation accidents in modern history.

Yeah, I reckon that's good enough.

The image of my daughter smiling in my dream is all the motivation I need as I drive away.

I have been in the air all my adult life. Now I've hung up my wings and joined the mortals.

# EPILOGUE

'The price of anything is the amount of life you exchange for it.' **Henry David Thoreau**

The story ends here, but the fight is far from over.

Qantas now uses the QF72 incident as a subject for recurrent training for pilots and cabin crew.

The surviving crew of QF72 are moving forward as best they can. Some are still at work, while some have left the company and its subsidiaries. I've mentioned a few of their names, with their permission. I hope you acknowledge them if your paths cross. Surviving the flight makes them some of the most experienced pilots and crew in Qantas; it also marks them as exceptional individuals.

I hope that by providing more detail, I've helped the QF72 passengers rest a bit easier. I salute them for their compassion and patience. Always keep your seatbelt fastened when you're seated, okay?

Part of this story is a warning: I survived a catastrophic failure of automation and am raising a red flag. Humankind continues to inject more automation into our daily lives, and we must question why.

In the past two years, there haven't been many automation failures in aviation, and I was starting to think that equipment and training were improving and awareness was increasing. Then, while this story is being written, there were two high-profile accidents involving the brand-new Boeing 737 MAX.

Lion Air JT610 was lost on 28 October 2018. A malfunctioning angle-of-attack sensor activated a stall protection mode not previously fitted to previous models of the 737 aircraft. The pilots fought as the confused protection system activated a nose-down control input. Ultimately, they lost control, and all 189 passengers and crew were killed.

In the new 737 MAX, a single angle-of-attack sensor supplies data to the MCAS (Maneuvering Characteristics Augmentation System) incorporated to enhance the aircraft's slow-speed stall characteristics through use of the pitch trim system. But what if that sensor malfunctions? It will still provide data to the MCAS and it could activate the protections. Has Boeing installed this system as a band-aid fix for poor aerodynamic performance? And how can pilots 'save the day' if they don't know about the system or how it can falsely activate and if there is no published procedure to deal with its abnormal operation? Sounds a lot like QF72 …

Then, there was the disaster that was Ethiopian Airlines flight 302. On 11 March 2019, shortly after takeoff, the pilots of ET302 followed procedures to regain control of a 737 MAX aircraft after the automated control system triggered several pitch-downs. All 157 people on board were killed.

The Preliminary Report on Ethiopian Air flight ET320 is disturbing. This 737 MAX malfunctioned in a similar fashion to Lion Air JT610 five months earlier.

According to the initial report, the MCAS activated erroneously, generating several pitch-downs that the pilots instinctively tried to arrest. Even though they followed recommended procedures to deactivate the trim system, MCAS had already introduced heavy nose-down tail trim that the pilots could not physically overcome.

Boeing Aircraft Corporation is now under scrutiny for their perceived failure to detail the existence of this protection system in the 737 MAX differences course and in the Flight Crew Operating Manual.

Investigations continue but, once again, the Achilles heel of sensor malfunction and data corruption has reared its ugly head in automated aircraft.

The Federal Aviation Administration is also under scrutiny for allowing aircraft manufacturers to 'self-certify' their aircraft with reduced regulator oversight. Is the move to self-regulation due to a lack of funding and resources, or are they protecting their turf in the competition for the supply of passenger aircraft globally?

Following the disaster of ET302, Ethiopian Airlines grounded their 737 MAX aircraft. Aviation regulators worldwide quickly followed suit, banning MAX operation within its respective airspace. It has since been revealed that critical aircraft monitoring systems that could have averted these accidents were sold as optional equipment. Those systems are now going to be fitted as standard. But it does seem that safety sometimes now comes at an additional price.

The investigation continues into these two tragedies.

We trust aircraft manufacturers to design and build safe aircraft. We expect the government regulators to certify, through rigorous and independent testing, that the aircraft are safe to carry passengers. And we trust that airline operators will respond appropriately in the name of safety, as the CEO of Ethiopian Air did in the wake of an aircraft accident of this magnitude.

QF72 is classified as an accident – the result of automation that failed, in a similar way to the automation on the two MAX flights, in which the pilots were powerless to stop its activation. The recent Boeing events are reminders that safety must remain the number-one priority for aircraft manufacturers, governments and airlines.

*

In late 2018, California Highway Patrol officers were challenged when a Tesla went zorching down the freeway at 70 miles per hour on 'autopilot'. A man was passed out in the driver's seat, so intoxicated that he was unresponsive to their sirens and lights as they gave chase. The officers improvised: one car blocked traffic behind the Tesla while the other raced ahead and took position in front of it. The police vehicle slowed, forcing the automation to react and eventually bring the vehicle to a stop. The driver was charged with drink-driving.

*

There's a price to pay when automation fails spectacularly.

Few have suffered more than Fuzzy Maiava. Fuzzy was employed by Qantas's wholly owned subsidiary Jetconnect, based in New Zealand, and was working under different conditions to the Australian-based crew. He's now on a disability pension.

The extreme forces generated during the pitch-downs inflicted serious injuries on Fuzzy. He suffered a compound fracture of his right fibula, meniscus tears on both knees, a fractured rotator cuff of his right shoulder and compression of the vertebrae in his lower spine, causing a permanent misalignment of his hips. His impact with the ceiling of the rear galley knocked him unconscious, and to this day, the resulting brain trauma causes unbalance and seizures.

He hasn't driven a car since that flight and will never be able to work again. Continued steroid injections are responsible for a weight gain of up to 168 kilograms.

Fuzzy began to isolate himself from his family and society. In 2012 his marriage broke down, and he tried to take his own life later in that year. He was then committed, by court order, to a mental hospital for eight weeks. The New Zealand compensation system ensures his medical treatment will continue, but he lives on the poverty line. He faces a life of medication for chronic pain and PTSD. The pain provides the triggers for nightmares.

On 7 October 2018, the ten-year anniversary of the QF72 accident, Fuzzy's daughter was stuck by a speeding motorist while walking home. She suffered severe injuries and continues to be treated and rehabilitated in an Auckland hospital.

Fuzzy has briefed me personally on these ten torturous years and wanted it included in the story. He's trying to move forward but his circumstances aren't fair.

I am honouring his courage and resolve by sharing this here.

I continue to manage PTSD and adjust to an earthbound life. It isn't easy but I'm actively trying to replace the bad memories and experiences with good ones. Sometimes I fly as a passenger; I've never been afraid to do that. I try to look out of the window as much as I can because I don't like being tethered to our planet. I miss being up there, and I will fly as necessary in my travels to new and distracting destinations.

I wonder if the pilots controlling my fate know how terribly complicated their job can become in these automated machines. I am all for enhanced safety and technology in aviation. But is it the goal of tech companies to use automation to replace the pilot someday, or further diminish their contribution in operating an aircraft? The QF72 accident is the worst-case scenario for a modern pilot and pilots need to be prepared.

In the recent 737 MAX accidents, the pilots were ambushed by failed automation. They were engaged in their own knife-fight for survival, but were overcome. I feel gutted by their loss; this shouldn't have happened. They didn't stand a chance, even as they followed the recommended procedures.

I'm an advocate for pilots, regardless of where they live or what aircraft they fly. Aviation remains the noblest of professions. But to be able to successfully do their jobs, pilots must be trained effectively and aircraft design must not restrain them from performing their function as the last line of defence. I salute the pilots of the 737 MAX accidents, along with passengers and crew, and hope their sacrifice was not in vain.

This book is my way of sharing knowledge and experiences, in the hope that they'll help others. Along with my crew members, I'm not a victim but a survivor, one of the few to have survived such a catastrophic event caused by faulty automation. And I have a voice. I hope my warnings are heard.

# ACKNOWLEDGEMENTS

'There aren't any great men. There are just great challenges that ordinary men like you and me are forced by circumstances to meet.' **Adm. William Halsey**

A hero is someone admired for their courage, bravery and their personal qualities, and one who has achieved or survived something challenging or life threatening. There were many heroes on 7 October 2008.

It is time to thank the many who came forward and helped the survivors of the QF72 accident.

The pilots and crew were heroes. Even though most were physically injured during the pitch-downs, they continued to function and to lead. Diana Casey, although travelling 'off duty', performed incredible acts of courage and initiative after the pitch-downs subsided. She is an amazing heroine and role model.

The passengers onboard QF72, also physically injured and in complete shock at the violence they had endured, remained calm and followed the instructions from the crew. Once we were safely on the ground, I witnessed many acts of compassion and support bestowed upon complete strangers by each and every passenger. We were safe but a long way from home, and I acknowledge each and every one of you for responding to the crew's direction and guidance, and for remaining patient while awaiting rescue. We were bound by the common purpose of our unique survival situation and the power of the human spirit was shining brightly in the face of this adverse circumstance. That bond continues.

The West Australian Police activated and coordinated the rescue operation from their command centre in Perth. It is a success story, worthy of a movie script, but I can only give you these words of gratitude and thanks for mobilising the support we needed at Learmonth. The constables on the ground at the airport provided us with strength and security, and we even shared some humour in the wake of this cataclysmic emergency.

Medical staff from Exmouth Regional Hospital were activated to assist the injured once we landed. How can so few provide assistance to so many? It was support of biblical proportions, plain and simple; also worthy of a movie script. Regional staff from other medical centres were also mobilised and airlifted in to Learmonth to assist the staff there. Thank you all.

The Royal Flying Doctor Service flew in and out of Learmonth on their flights of mercy. I was awestruck as I watched them arrive and depart as the day turned into night, carrying their precious cargo of the seriously injured onward to larger medical facilities. I know all the crews flew excessive tours of duty that day in support of the rescue mission, and the RFDS Base at Jandakot was buzzing with activity and support. I salute you.

There were airport staff and volunteers at Learmonth who assisted and supported the shell-shocked passengers as best they could. Somehow, food and drink materialised out of thin air to sustain the survivors and I know it was the initiative of these individuals that made this possible. The ham and cheese sandwiches were the best I've ever had, even to this day. Thank you.

Melbourne Air Traffic Control cleared the airspace for our arrival into Learmonth. The controller on duty was of exceptional skill and I have benefited from his calm and precise expertise on a previous occasion. Thank you.

The pilots and crew of the two rescue aircraft also performed an extended tour of duty in the process of airlifting the survivors out of Learmonth. Thrust into this unprepared role of support, the crews stepped up and provided safety and compassion to us and delivered us safely to Perth.

Medical staff at Sir Charles Gairdner Hospital were activated to provide a triage service at Perth Airport. I observed the compassion and support you bestowed on the surviving passengers in the arrivals terminal. Thank you for the support you provided at the airport and in the Intensive Care Unit at the hospital.

Captains John Killingback and Phil Paterson were deployed to Perth to assist the pilots once we arrived and to coordinate our return to Sydney. You were the cavalry I needed, joining the battle at the critical point, after all my energy and ammunition had been expended. You saved my scalp.

Captain Steve Anderson, the Welfare Officer at our pilots' association, was the first to contact me once we landed safely at

Learmonth. His Aussie humour and 'tender words of concern' was the tonic I needed to deal with the many challenges that confronted thc survivors as we waited for rescue.

Mr Adam Lloyd is a saint and a hero in my mind. He is no longer employed at Qantas and it is the organisation's loss that he has departed now. He planned and implemented the repatriation of my daughter to Sydney and provided emotional support to her when she needed it most. He was a pillar of compassion and strength as we all struggled to cope with the aftermath of the accident. Thank you Adam.

My friends are a very stout and exceptional group who live around the world. They continue to support me in the face of PTSD, even though it is difficult for all of us to manage. The good news is that most of you are still my friends; some have fallen by the wayside. Thank you for those who are standing next to me as I battle to rebuild my life.

This story is dedicated to my daughter. She doesn't realise that she has kept me alive. She has had to watch her father suffer over the years since the accident flight, even though he is supposed to be a hero. Her love and devotion cannot be described in words, but all should know that she is my heroine.

This story is real. It happened. I have acknowledged those individuals and organisations who remembered their humanity and extended their compassion to help those in need. I am a human being and I want to help, too.

PTSD is a silent killer and I have been examined and diagnosed with it. It doesn't go away and it places the sufferer on a slippery slope to potential ruin and possibly death. Identifying with a sufferer is difficult. For many, though, the damage is plain to see: prosthetic limbs and scars mark the trail of their tortuous journeys. For others, they appear whole and undamaged but their behaviour has changed forever. Depression and PTSD are inseparable villains who visit those who have been damaged, tortured and abused by forces outside of their control.

There are charities and organisations that exist to assist PTSD sufferers. I want to support them as they work to support others. A proportion of the royalties of this book will be donated to these groups. I have a list already in mind and your purchase will benefit others. I hope my honesty in relating my struggles with PTSD may help others to find the courage to fight it, rather than succumb to it.

I wish this had never happened but I've got no time to cry over spilled milk.

In Memorium: John 'Breeze' Martin (USN, Ret.)

# BIBLIOGRAPHY


**Aviation Safety – General**

Air Accidents Investigation Branch (UK), 'Report of Airbus A330 C-GGWD and Airbus A340 TC-JDN, 2 October 2000', www.gov.uk/aaib-reports, December 2014.

Australian Transport Safety Bureau, 'Air Data System Failure Involving Airbus A330-243, A6-EYJ, 21 November 2013', atsb.gov.au, May 2016.

Australian Transport Safety Bureau, 'Erratic Speed Indication, Boeing 787-8, VH-VKE, 21 December 2015', atsb.gov.au, June 2017.

Bureau d'Enquêtes et d'Analyses pour la Sécurité de l'Aviation Civile (BEA), 'AF447, F-GZCP, Final Report on Accident of 1 June, 2009', www.bea.aero, July 2102.

Bureau d'Enquêtes et d'Analyses pour la Sécurité de l'Aviation Civile (BEA), 'Air France 471, F-GLZU, Serious Incident on 22 July 2011', www.bea.aero, May 2012.

Bureau d'Enquêtes et d'Analyses pour la Sécurité de l'Aviation Civile (BEA), 'Final Report, Germanwings, D-AIPX, 24 March 2015', www.bea.aero

Comisión de Investigación de Accidentes e Incidentes (CIAIAC), 'Technical Report A006/2001, Accident of Aircraft Airbus A320-214, Registration EC-HKJ, at Bilbao Airport, 7 February 2001', November 2006.

Croft, John, 'German Investigators Cast Wider Net For Frozen AOA Sensors in Pamplona Dive Incident', *Aviation Daily*, March 25 2015.

Dieusaer, Tom, 'Airbus 330 design flaws under scrutiny in the AF447 court case', medium.com,

Federal Aviation Authority, 'Air Carrier Training/ Enhanced Pilot Training and Qualification Requirements', faa.gov, December 2017.

Federal Aviation Authority, 'Stall and Stick Pusher Training', FAA Advisory Circular AC 120-109, 2012.

Flight Safety Foundation, 'Lufthansa LH-1829, D-AIDP, 5 Nov 2014', aviation-safety.net

Hepher, Tim, 'A400M Probe Focuses on Impact of Accidental Data Wipe', Reuters, 10 June 2015.

Hepher, Tim, 'Airbus Knew of Software Vulnerability Before A400M Crash', Reuters, 8 November 2017.

Honeywell Aerospace, 'Honeywell's Inertial Navigation System Becomes Standard Equipment on Airbus', Honeywell Press Release, honeywell.com, 30 April 2010.

Hradecky, Simon, 'Air France A332 over Tanzania on Feb 12th 2012, Severe Turbulence and Temporary Loss of Control', *The Aviation Herald*, www.avherald.com, July 2014

International Civil Aviation Organisation (ICAO), 'Manual on Aeroplane Upset Prevention and Recovery Training', Doc 10011, First Edition, 2014.

Komite Nasional Keselamatan Transportasi (Indonesia), 'Aircraft Accident Investigation Report, Airbus A320-216, PK-AXC', December 2015.

Krahe, Chris, 'The Iron Bird', *Airbus Technical Digest: Fast 24*, May 1999

Lacagnina, Mark, 'Ice Blocks A330 Pitot Probes', flightsafety.org, September 2011.

Langewiesche , William, 'Should Airplanes be Flying Themselves', *Vanity Fair*, October, 2014.

Learmount, David, 'Details of A330 Angle-Of-Attack Sensor Incident Emerge', *Flight Global*, Dec 2012.

Levin, Alan, 'Lion Air Crash Sheds Light on Cockpit Automation's Dark Side', Bloomberg, November 2018.

NASA Safety Center, 'What's Happening? The loss of Air France 447', August 2012.

Reverbel, Pierre-Henri, 'Blocked AoA Probes/ Temporary Procedure for Aircraft Equipped with Conic Cover Plates', Airbus via scribd.com, December 2012.

SKYbrary, 'A320 Vicinity of Perpignan France 2008', www.skybrary.aero

SKYbrary, 'Accident and Serious Incidents: Loss of Control/ Non-Normal FBW Flight Control Status', www.skybrary.aero

Stock, Stephen and Putnam, Julie, 'FAA Report: Pilots Addicted to Automation', NBC Bay Area, 22 November 2013.

Thompson, Mark, 'A Past Flight May Offer Clues to Air France 447', time.com, June 2009.

Thurber, Matt, 'Upset Training in the Age of Fly BY Wire', *Business Aviation*, May 2016.

Traufetter, Gerald, 'Doomed Flight AF 447/Questions Raised About Airbus Automated Control System', Spiegel Online, May 2011.

Traufetter, Gerald, 'The Computer vs. the Captain, Will Increasing Automation Make Jets Less Safe?', Speigel Online, July 31, 2009.

Wainwright, William, 'Airplane Upset Recovery' *Airbus Technical Digest: Fast 24*, May 1999

Wise, Jeff, 'What Really Happened Aboard Air France 447', *Popular Mechanics*, 7 December 2011.

Zalagenaite, Zivile, 'Blame the Pilot! The Case of Air France 447', aerotime.aero, November 2017.

**Automation**

Ackerman, Evan, 'An Autonomous Passenger Drone Seems Like a Terrible Idea', *IEEE Spectrum*, February 2107.

Bromey, Nathan, 'Self Driving Cars Could Kill Hundreds But Save Tens of Thousands, Toyota Executive Says', *USA Today*, 29 March 2018.

Charette, Robert N. 'Automated to Death', *IEEE Spectrum*, December 2009.

Elish, Madelaine and Hwang, Tom, 'Praise the Machine! Punish the Human! The Contradictory History of Accountability in Automated Aviation', Madeleine Elish and Tim Hwang, Data & Society Research Institute, 24 February 2015.

Elish, Madelaine, 'Letting Autopilot Off the Hook: Why do we blame humans when automation fails?', Slate.com, 16 June 2016.

Greenemeier, Larry, 'Deadly Tesla Crash Exposes Confusion Over Automated Driving', *Scientific American*, July 2016.

Hiraki, Hiraki and Warnink, Mike, 'Automation Bias and Surprise', Cockpit Aviation Fact Sheet, Luchtvaartfeiten.nl, February 2016.

Marshall, Aarian, 'Why UBER's Self-Driving Crash Is Confusing for Humans', WIRED.com, March 31, 2018.

Merchant, Brian, 'The Deadly Recklessness of the Self-Driving Car Industry', Gizmodo.com, December 2018.

Naughton, Keith and Beene, Ryan, 'The Self-Driving UBER in Fatal Crash Didn't Have a Vision Problem', Bloomberg, 22 March 2018.

Wiener, Earl L. and Curry, Renwick, 'Flight Deck Automation, Promises and Problems', NASA Technical Memorandum 81206, NASA, June 1980.

Williams, Elliot, 'Fatalities vs False Positives: The Lessons from the Tesla and UBER Crasher', hackaday.com, June 2018.

**PTSD & Pilot Mental Health**

Brooksby, Scott and Alexander, Brian J. 'What Can Be Done About Pilot Depression, Suicide, and Other Flight Crew Mental Health Issues?', americanbar.org, November 2017.

Burke, Susan, 'When Your Mental State Cries MAYDAY', *Air Line Pilot*, August 2007.

Grossman, David, 'Combat Stress Response & Tactical Breathing (Modified for the Combat Aviator)', *On Combat*, summary by Rocky Apollo, goflightmedicine.com, May 2014.

Jeffreys, Matt MD, 'Clinician's Guide to Medications for PTSD', Veterans Affairs and National Center for PTSD, 2017.

Kulish, Nichols and Clark, Nicola, 'Germanwings Crash Exposes History of Denial on Risk of Pilot Suicide', *New York Times*, 18 April 2105.

Notyce, Corina, 'Tips to Help Kick Start Your Emotions', Defence Center of Excellence for Psychological Health and Traumatic Brain Injury, February 2012.

Nutt, D.J., 'The Psychobiology of Post Traumatic Stress Disorder', *Journal of Clinical Psychiatry*, 2000.

PTSDUnited.org, 'PTSD Statistics'.

Royal College of Psychiatrists, 'Post Traumatic Stress Disorder', rcpsych.ac.uk, April 2013.

Southwest Airlines Pilots' Association, 'Pilots Helping Pilots: Mental Health for Pilots and Their Families', swapafamily.org

**Military**

*CV NATOPS Manual*, NAVAIR 00-80T-105, Naval Air Systems Command, 21 October, 1999.

*NATOPS Landing Signal Officer Manual*, NAVAIR 00-80T-104, May 2009.

'Out of Control Flight/ T-45 Strike', Flight Training Instruction, CNATRA P-1216 (Rev. 5-12), Naval Air Training Command, 2012.

**QF72**

Albright, James, 'QF 72 Case Study', code7700.com

Australian Transport Safety Bureau, 'Investigation AO-2008-070, VH-QPA', Preliminary, Interim, Final Reports of QF72, atsb.gov.au.

Aviation International News, 'Fly By Wire Failure' and 'Tales From the Flight Deck', Episodes 7 and 8, podcast, ainonline.com

Carroll, Joanne, 'Hurt Kiwi Fighting On', *The New Zealand Herald*, 24 February 2013.

Coulthart, Ross 'Airbus Transcript', Yahoo 7, 20 October 2009.

'Mid-Air Terror as QF72 Plunges', News.Com.au, 31 August 2009.

National Geographic, *Air Crash Investigation/Mayday*, 'Free Fall', Season 18 Episode 7.

O'Sullivan, Matt, 'The Untold Story of QF72: What Happens When 'Psycho' Automation Leaves Pilots Powerless?', *The Sydney Morning Herald*, 12 May 2017.

Ramachandran, Arjun, 'Ex-Fighter Pilot Saved QF72', *Sydney Morning Herald*, October 15, 2008.

Squair, Matthew, 'QF72 and Sense-Making in the Cockpit', CriticalUncertanties.com, 13 May 2017

Thompson, Mark, 'A Past Flight May Offer Clues to AF 447', *TIME*, 3 June 2009.